# POSTHARVEST

## About the authors

**Dr Ron Wills** is Professor in the Department of Food Technology at the University of Newcastle, Australia. He has had 30 years' experience in many aspects of postharvest horticulture and has published more than 230 papers. He has held government, university and industry positions in Australia and New Zealand and has consulted governments and international agencies on horticultural development projects throughout the Asia-Pacific region.

**Dr Barry McGlasson** was Senior Principal Research Scientist in the CSIRO Division of Horticulture and is now a Fellow in the Faculty of Horticulture at the University of Western Sydney (Hawkesbury), Australia, where he has co-ordinated postharvest studies within the undergraduate Horticultural Science program.

**Dr Doug Graham** is formerly Head of the Food Research Laboratory in the CSIRO Division of Food Science and Technology, Australia. He has had extensive experience in the biochemistry and physiology of horticultural crops.

**Dr Daryl Joyce** is Senior Lecturer in Postharvest Horticulture at Gatton College, The University of Queensland, Australia, where he teaches postharvest and other plant science subjects to undergraduate students. Dr Joyce gained considerable experience in postharvest horticulture as a Postdoctoral Research Associate at the University of California, Davis, as well as in several teaching and research positions in Austra

Ron WILLS + Barry McGLASSON + Doug GRAHAM + Daryl JOYCE

# 4TH EDITION

# Postharvest

## An Introduction to the Physiology & Handling
## OF FRUIT, VEGETABLES & ORNAMENTALS

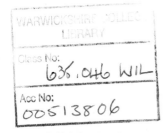
*Published in Australia, New Zealand, Papua New Guinea and Oceania by*
University of New South Wales Press Ltd
University of New South Wales
Sydney 2052 Australia

*and in the rest of the world by*
CAB INTERNATIONAL
Wallingford Oxon OX10 8DE UK
Tel + 44 (0) 1491 832111  Fax + 44 (0) 1491 833508
Email <cabi@cabi.org>

*and* CAB INTERNATIONAL
198 Madison Avenue
New York, NY 10016-4314  USA
Tel + 1 212 726 6490  Fax + 1 212 686 7993
Email <cabi-nao@cabi.org>

National Library of Australia
Cataloguing-in-Publication entry:

Postharvest: an introduction to the physiology and handling
of fruit, vegetables and ornamentals.

4th ed.
Includes index.
ISBN 0 86840 560 4.

1. Vegetable—Postharvest physiology. 2. Vegetable—
Postharvest technology. 3. Fruit—Postharvest physiology.
4. Fruit—Postharvest technology. 5. Plants, Ornamental—
Postharvest physiology. 6. Plants, Ornamental—Postharvest
technology. I. Wills, R.B.H. (Ronald Baden Howe).

635.046

A catalogue record for this book is available from the British Library.
A catalogue record for this book is available from the Library of Congress,
Washington, DC, USA.

ISBN 0 86840 560 4 (UNSW Press)
ISBN 0 85199 264 1 (CABI)

Printed by Hyde Park Press, Adelaide, South Australia

# CONTENTS

PREFACE *vii*

ACKNOWLEDGMENTS *viii*

LIST OF ILLUSTRATIONS *ix*

1  Introduction  *1*
2  Structure and composition  *15*
3  Physiology and biochemistry  *33*
4  Effects of temperature  *60*
5  Water loss and humidity  *77*
6  Storage atmosphere  *97*
7  Technology of storage  *113*
8  Physiological disorders  *130*
9  Pathology  *144*
10  Evaluation and management of quality  *159*
11  Preparation for market  *188*
12  Packaging  *214*
13  Commodity storage recommendations  *230*

APPENDICES

I  List of abbreviations  *241*
II  Glossary of plant names  *243*
III  Temperature and humidity measurement  *246*
IV  Gas analysis  *254*

Index  *257*

# PREFACE

Since the conception of this book some 17 years ago there has been a growing appreciation of the importance of the correct handling and storage of fresh fruit and vegetables, and accredited coursework in this area of study is now included in the curricula of tertiary teaching institutions throughout the world that are associated with horticulture or food technology. The rapid expansion of the ornamental industry and its development as a strong international trading commodity sector has now given ornamentals recognition as an important component of postharvest horticulture. The current edition of this book, while retaining the essential organisation and spirit of previous editions, has been substantially updated to reflect modern industry practices and technology, and now includes material on ornamentals.

This book initially examines the background information on horticultural produce, which is discussed in terms of structure, composition, physiology and biochemistry. The effect of environmental conditions such as temperature, relative humidity and gas atmospheres are examined for their effects on postharvest behaviour, along with discussion of the technology available to control these conditions and its commercial application, including during storage. Then follows discussion of physiological disorders, pathology, microbial wastage and other factors leading to loss of produce quality. A discussion of quality evaluation and quality management integrates the technical and market requirements into formal operating systems that ensure quality and profitability for commercial entities. The final chapters examine the logistics and infrastructure required to transfer produce from the farm to the end user, and details recommendations for optimum postharvest handling of individual produce.

The text is intended for use in tertiary courses at technical colleges and universities and as a useful guide for individuals who may be employed as technologists by farming companies, transport organisations and retailing organisations, as well as for packing house managers, cool storage operators, nutritionists and advisers to governments who are involved with horticultural produce. It should also be of value to the concerned consumer.

## Acknowledgments

The present authors acknowledge with gratitude the contributions made by former authors Dr Terry H. Lee, now Vice President (Technical Services & Research and Development) E&J Gallo Winery, Modesto, CA, USA, and the late Mr Eric G. Hall, formerly a Principal Research Scientist at the CSIRO Division of Food Research, Sydney, Australia, to earlier editions of this book, some of which material has been incorporated with revisions into this edition.

The authors also gratefully acknowledge contributions to the revision of specific sections of the book by the following colleagues: Mr Brian B. Beattie, Postharvest Horticulturist, New South Wales Department of Agriculture, Australia (postharvest systems and quality assurance); Ms Clare Hamilton-Bate, Consultant, Meyers Strategy Group P/L, New South Wales, Australia (retail systems and quality assurance); and Dr Brian L. Wild, Senior Research Horticulturist, New South Wales Department of Agriculture, Australia (postharvest pathology). The assistance of Ms Jennifer Bower (postgraduate student, University of Western Sydney, Hawkesbury, New South Wales, Australia) in the preparation of the illustrations is gratefully acknowledged.

## List of illustrations

**Figure 2.1** Derivation of some fruits from plant tissue 16

**Figure 2.2** Derivation of some vegetables from plant tissue 18

**Figure 2.3** Examples of the variations in the structure of flowers 19

**Figure 2.4** Diagrammatic representation of a plant cell 20

**Figure 3.1** Growth, respiration and ethylene production patterns of climacteric and non-climacteric plant organs 34

**Figure 3.2** Physiochemical changes that occur during ripening of harvested tomatoes at 20°C 35

**Figure 3.3** Physiochemical changes that occur during ripening of the Cavendish banana (variety Williams) 37

**Figure 3.4** Respiratory patterns of some harvested climacteric fruits 38

**Figure 3.5** Effects of applied ethylene on respiration of climacteric and non-climacteric fruits 41

**Figure 3.6** Starch and sucrose degradation to form the 6-carbon sugars, glucose and fructose, for oxidation via glycolysis 48

**Figure 3.7** A simplified scheme for aerobic respiration of carbohydrate reserves in plants via glycolysis and the TCA cycle 48

**Figure 3.8** NADH and FADH2 are oxidised via the electron transportchain to produce water and ATP form ADP 49

**Figure 3.9** Oxidative Pentose Phosphate Pathway 51

**Figure 3.10** Some pathways for the degradation of chlorophyll 55

**Figure 4.1** Responses of non-chilling-sensitive and chilling-sensitive produce to temperature 60

**Figure 4.2** Illustration of a simple hypothetical Q1° relationship in plants and the effect on quality retention (i.e. time of shelf life) 63

**Figure 4.3** Cooling system 1 illustrates the ideal curve to achieve seven-eighths cooling (three half-cooling times, or 12.5% of the original temperature) in 9 hours 69

**Figure 4.4** Air flow in forced-air (pressure) cooling 72

**Figure 4.5** Vacuum cooler designed for cooling two pallets of fresh produce at a time 74

**Figure 5.1** Relationship between weight loss and green life of mango fruit at 20°C 77

**Figure 5.2** Loss of ascorbic acid (vitamin C) during storage of exposed (open) and wrapped in plastic (closed) Brassica juncea leaves at 24–28°C 78

**Figure 5.3** A Hofler diagram derived from analysis of the Y isotherm and showing the relation between relative water content (RWC) and water potential (YMP), osmotic potential (YOP) and turgor potential (YTP) 79

**Figure 5.4** Change in the radius of attached mango fruit in relation to vapour pressure deficit (VPI, kPa) over a 3-day period 80

**Figure 5.5** Growth in colony diameter of Botrytis cinerea on a solid medium after 3 days at various equilibrium relative humidities (ERH) at 20°C 82

**Figure 5.6** Simplified psychometric chart 83

**Figure 5.7** Relation between equilibrium relative humidity and water potential of plant tissues at 0°C 85

**Figure 5.8** Cross-section of a leaf showing the surface (cuticle), stoma, internal structures and the network of intercellular spaces 87

**Figure 5.9** Effect of vapour pressure difference (deficit) on weight loss by the apple cultivars Golden Delicious (upper curve) and Jonathan (lower curve) 89

**Figure 6.1** Relative tolerance of fruit and vegetables to elevated carbon dioxide and reduced oxygen concen-trations at storage temperatures 101

**Figure 6.2** System for removal of ethylene with atomic oxygen generated by ultraviolet radiation 110

**Figure 7.1** Cassava storage clamp 115

**Figure 7.2** Basic component parts of a mechanical refrigeration plant 116

**Figure 7.3** Modern cool room being constructed of metal clad panels insulated with polystyrene insulation 118

Figure 7.4 A modern cool room in which the cool air is blown through the coils by fans mounted at the back of the coils (force draft cooler) 119

Figure 7.5 Pressure swing adsorption machine 121

Figure 7.6 Hollow fibre membrane system 122

Figure 7.7 Integral-insulated shipping container with a built-in refrigeration unit 128

Figure 7.8 A porthole refrigerated container must be coupled to either a clip-on refrigeration unit or to a source of cool air generated by a land-based or a shipboard refrigeration unit 128

Figure 7.9 To ensure effective air circulation in a refrigerated road vehicle there must be an air delivery chute, ribs on the doors and walls, channels or pallets on the floor and a return bulkhead 128

Figure 8.1 Chilling injury in avocado fruit appears as browning of the mesocarp due to the breakdown of cell compartmentalisation and the action of polyphenol oxidases to produce tannins 132

Figure 8.2 Storage life at various temperatures of produce 133

Figure 8.3 Time sequence of events leading to chilling injury 134

Figure 8.4 Illustrations of examples of non-pathogenic diseases colour plate

Figure 8.5 Equipment for the flood application of diphenylamine to bulk boxes of apples and pears before cool storage 140

Figure 9.1 Illustrations of pathological diseases colour plate

Figure 9.2 Flood application of Thiabendazole to oranges colour plate

Figure 9.3 Wax emulsion being sprayed onto oranges colour plate

Figure 9.4 A non-pathogenic fungus, Paecilomyces sp., engulfing Geotrichum candidum colour plate

Figure 10.1 Modern machinery for sizing, colour sorting and packing fruit 164

Figure 10.2 Ripening scale of 'Cavendish' banana (Musa acuminata, var Williams) colour plate

Figure 10.3 Commercial maturity in relation to developmental stages of the plant 169

Figure 10.4 Tomato colour chart colour plate

Figure 10.5 Diagram of the geometry used to make transmission measurements of fruit 175

Figure 10.6 Effegi and Magness-Taylor fruit pressure testers 176

Figure 10.7 Variation in soluble solids/acid ratio in juice from Washington Navel and Valencia Late oranges grown on trifoliata rootstocks 177

Figure 10.8 Electromagnetic spectrum illustrating the wavelengths of radiation that can be used for the non-destructive detection of changes in the structure and composition of plant tissues 178

Figure 11.1 Establishment of effective treatments for the disinfestation of produce 198

Figure 11.2 Vapour heat apparatus 200

Figure 11.3 Controlled ripening room for bananas 207

Figure 11.4 Trickle system for adding ethylene to a ripening room 208

Figure 12.1 Typical bamboo basket used throughout Southeast Asia for the handling and transportation of produce 215

Figure 12.2 Shoulder transport of taro in bamboo baskets in Indonesia 215

Figure 12.3 (A) Produce in fibreboard cartons (B) Bulk bins made from fibreboard 217

Figure 12.4 (A) Boxes made of foamed polystyrene (B) Onions in plastic mesh bags; carrots in large plastic bags; potatoes in woven jute and plastic bags 218

Figure 12.5 Handling pallet loads of packaged produce in a room used for temporary cool storage 221

Figure 12.6 (A) Self-service displays of prepackaged and loose fruit and vegetables in a modern supermarket (B) Prepacked whole and fresh-cut vegetables in a refrigerated display cabinet 225

# 1
# INTRODUCTION

## IMPORTANCE OF FRUIT AND VEGETABLES AS FOOD

Fresh fruit and vegetables have been a part of the human diet since the dawn of history. However, their full nutritional importance has only been recognised in recent times. Western societies have tended to value foods from animal sources more highly. On the other hand, other societies with diets that are largely or totally vegetarian for religious or economic reasons have had greater dependence on fruit and vegetables for survival. With the assistance of modern nutritional science, the image of fruit and vegetables has now risen considerably, and health professionals — particularly in developed countries — are actively recommending increased consumption of fruit and vegetables and restricted consumption of animal foods.

The nutritional value of some fruit and vegetables was recognised in the early 17th century in England. One example is the ability of citrus fruit to cure the disease scurvy, which was widespread among naval personnel. While individual captains took advantage of this knowledge to maintain the health of crews on long voyages, it was not until the late 18th century that the British Royal Navy issued a regular ration of lime juice to all sailors, leading to the term "limeys" applied to British sailors.

The discovery of ascorbic acid (vitamin C) as the ingredient responsible for the prevention of scurvy did not occur until the 1930s. It has since been shown to have a range of beneficial effects related to wound healing and as an antioxidant. There is now also considerable speculation as to its possible action as an anti-viral and anti-cancer agent. Dietary sources of vitamin C are essential as humans are unable to synthesise it. All fruit and vegetables contain vitamin C, and as a group are the major dietary source supplying about 95 per cent of body requirements in virtually all countries.

Specific fruit and vegetables are also excellent sources of the provitamin A carotenoids, which are essential for the maintenance of ocular health, as well as folic acid, which prevents certain anaemias. FAO and WHO have active programs promoting the use of home vegetable gardens as an inexpensive and readily available approach to combat vitamin deficiency diseases in less developed regions.

The rise in the nutritional importance of fruit and vegetables has been stimulated by a range of degenerative diseases prevalent in sedentary, affluent societies, particularly in Western countries. Many of these diseases are attributed, in part at least, to a diet inappropriate for the modern urban lifestyle. Concerns over obesity and coronary heart disease have led to the promotion of reduced levels of fat in the diet, while dietary fibre is considered to be beneficial in reducing or preventing a raft of medical conditions including appendicitis, colonic and rectal cancers, constipation, diabetes, diverticulitis, gallstones, haemorrhoids, hiatus hernia and varicose veins. Fruit and vegetables are generally low in fat and high in dietary fibre and are thus promoted as a substitute for animal-based foods and highly refined plant-based foods.

The status of fresh fruit and vegetables has also benefited from an international trend towards fresh natural foods, which are perceived to be superior to processed foods and contain less chemical additives. This community perception has, however, placed additional pressures on the horticultural industry to retain its fresh natural image by minimising the use of synthetic chemicals during production and postharvest handling. Notwithstanding nutritional status and fresh natural appeal, the attraction of fruit and vegetables for many consumers is from the sensory stimulation they impart during usage. Fruit and vegetables provide variety in the diet through differences in colour, shape, taste, aroma and texture and the diversity of these attributes between individual produce distinguishes fruit and vegetables from the other major food groups of grain, meat and dairy products. The sensory appeal of fruit and vegetables is not confined to consumption, but also has market value. Diversity in colour and shape is used to great effect by traders in product displays to attract potential purchasers, and chefs have traditionally used fruit and vegetables to enhance the attractiveness of prepared dishes or table presentations. The use of parsley and similar herbs to adorn meat displays is widely used throughout the Western world while fruit and vegetable carvings as table ornaments have become an art form in countries such as Thailand.

While the nutritional composition of ornamental horticultural crops is generally inconsequential to consumers, flowers are increasingly being included in prepared mixed salads, and therefore do make a limited contribution to the diet. However, the principal contribution of ornamentals can be considered as food for the mind, through their provision of senses of pleasure and serenity derived from the colour, shape and aroma of individual species. Apart from the more traditional home uses of garden plants and cut flowers, foliage and flowering plants are being increasingly used in interiorscapes in commercial premises, including offices, hotels and restaurants. The importance of ornamentals in the cultural life of societies should not be underestimated. Considerable commercial opportunites arise from their role in ceremonies such as weddings and funerals, in conveying messages on special occasions such as Mother's Day and Valentine's Day, as decorations in parades and rallies, and in art and creative pastimes, as reflected in the growth of ikebana schools. Ornamentals are given national status, with flowers being a symbol of state for most countries.

# HORTICULTURAL PRODUCTION STATISTICS

......

## Fruit and Vegetables

Production of fruit and vegetables worldwide has been increasing over many years, partly in response to a rising world population but also due to rising living standards in most countries and active promotion by government health agencies of fruit and vegetable consumption. Table 1.1 shows that total world production of fruit and vegetables has increased by about 50 per cent in the 24 year period from 1970–94. The fastest growing regions with 100 per cent increases in production were Asia, which now produces half the world total, and Africa. Europe is the only region to show a decrease, with a 15 per cent decline in production over the 24 years. Table 1.1 also shows production data for produce

TABLE 1.1 WORLD PRODUCTION OF FRUIT AND VEGETABLES (MILLION TONNES)

|  | ALL PRODUCE | | | | NON-ROOT CROPS | | | |
|---|---|---|---|---|---|---|---|---|
|  | 1970 | 1980 | 1990 | 1994 | 1970 | 1980 | 1990 | 1994 |
| World | 1010 | 1215 | 1389 | 1457 | 498 | 664 | 814 | 873 |
| Africa | 112 | 133 | 192 | 203 | 43 | 58 | 83 | 86 |
| Asia | 344 | 503 | 612 | 673 | 191 | 282 | 380 | 432 |
| South/Central America | 103 | 121 | 154 | 155 | 54 | 72 | 103 | 105 |
| USA/Canada | 65 | 80 | 81 | 93 | 47 | 56 | 60 | 66 |
| Europe | 378 | 371 | 344 | 319 | 158 | 185 | 183 | 180 |
| Oceania | 7 | 9 | 11 | 12 | 5 | 6 | 7 | 8 |

SOURCE FAO (1995) FAO yearbook: Production 1994, vol. 48, FAO, Rome.

TABLE 1.2  MAJOR NATIONAL PRODUCERS OF FRUIT AND VEGETABLES IN 1994 (MILLION TONNES)

| | ALL PRODUCE | | ROOT CROPS | | OTHER VEGETABLES | | FRUIT | |
|---|---|---|---|---|---|---|---|---|
| | COUNTRY | TONNAGE | COUNTRY | TONNAGE | COUNTRY | TONNAGE | COUNTRY | TONNAGE |
| 1 | China | 316 | China | 150 | China | 129 | China | 37 |
| 2 | India | 113 | Nigeria | 44 | India | 65 | India | 33 |
| 3 | USA | 86 | Russia | 34 | USA | 36 | Brazil | 32 |
| 4 | Brazil | 65 | Brazil | 27 | Turkey | 19 | USA | 29 |
| 5 | Nigeria | 58 | Poland | 23 | Japan | 14 | Italy | 18 |
| 6 | Russia | 47 | USA | 21 | Spain | 11 | Spain | 12 |
| 7 | Italy | 34 | India | 21 | Russia | 10 | France | 11 |
| 8 | Turkey | 33 | Zaire | 20 | Iran | 10 | Turkey | 10 |
| 9 | Indonesia | 30 | Thailand | 19 | Korea, Rep | 10 | Mexico | 10 |
| 10 | Poland | 30 | Indonesia | 18 | Egypt | 9 | Iran | 9 |

SOURCE  FAO (1995) *FAO yearbook: Production, 1994*, vol. 48, FAO, Rome.

excluding root vegetables, which now comprise 40 per cent of total production, down from 50 per cent in 1970. The increase in fruits and non-root vegetables over the period was 75 per cent, with increases of 125 per cent in Asia and 100 per cent in Africa and South/Central America. Europe showed a small increase from 1970–80 but no increase in the past 14 years. The fall in total European production over the period is thus due to a decreased tonnage of root vegetables.

Table 1.2 lists countries that are the top 10 producers of fruit and vegetables. As might be expected, the lists for total production and sub-commodity groupings are dominated by the most populous countries — China and India — which between them generate 30 per cent of total world production. Production in both countries grew by 30 per cent over the past 10 years, which was faster than the rest of the world. While much of this increased production would be consumed domestically, considerable efforts, particularly in China, have been directed to expanding the export trade. The export drive explains much of the increase in China's fruit production, which increased by 25 million tonnes or 210 per cent over the past 10 years. Many of the other major producers such as Brazil, Turkey, Thailand and Mexico have increased production through a national export trade effort targeting fruit and vegetables. Traditional European exporters such as Italy, France and Spain have shown little if any growth in the past 10 years.

The international trade in fruit and vegetables has been growing rapidly, with the total value of trade increasing from about US$25 million in 1984 to US$50 million in 1993. Table 1.3 lists the major countries involved in import and export. The USA is the dominant country and is directly associated with about 25 per cent of world trade. Germany, France, Italy, the Netherlands and Spain are also prominent in both import and export trade. Of the developing countries, China, Mexico, Turkey and Thailand are important exporters.

## Ornamentals

Ornamentals comprise an important sector of horticulture. Ornamental crops include cut flowers and foliage, flowering and foliage pot plants, bedding plants, and containerised shrubs and trees. Worldwide consumption of such crops was around US$25 billion per annum in 1990 (de Boon, cited by Behe 1993). Compared with fruit and vegetables, ornamentals have a substantially higher multiplier value in terms of secondary industries such as processing versus retailing, plant hire, and interiorscaping; Stephenson (1989) has suggested relative multipliers of 0.48, 0.36 and 7.97, respectively. Due to the many allied secondary industries that rely on fresh ornamentals, postharvest (or more correctly, postproduction in many instances) handling is a critical issue.

TABLE 1.3  VALUE OF INTERNATIONAL TRADE IN
FRUIT AND VEGETABLES IN 1993 (US $MILLION)

| COUNTRY | IMPORTS | COUNTRY | EXPORTS |
|---|---|---|---|
| Germany | 9110 | USA | 6370 |
| USA | 6550 | Spain | 5560 |
| Japan | 4870 | Netherlands | 5260 |
| France | 4800 | Italy | 3750 |
| UK | 4550 | France | 2900 |
| Netherlands | 3650 | China | 2640 |
| Canada | 2670 | Mexico | 1850 |
| Italy | 2060 | Turkey | 1670 |
| Hong Kong | 1410 | Thailand | 1610 |
| Spain | 1170 | Germany | 1490 |

Total international trade of US $50 000 million

SOURCE  FAO (1994) *FAO yearbook: Trade, 1993*, vol. 47, FAO, Rome.

The world trade in ornamentals is dominated by western Europe, most notably by The Netherlands. Rose and Ficus headed the top 10 cut flower and pot plant lists, respectively, of lines traded in 1994 through the Dutch auction system (Table 1.4). In 1994, the top six cut flower export countries were, in descending order: The Netherlands, Colombia, Italy, Israel, Spain and Kenya (Anon, 1995). The corresponding top five pot plant exporter countries were The Netherlands, Denmark, Belgium, Germany and Italy.

In terms of per capita consumption of cut flowers, western European countries from Scandinavia to the Mediterranean feature in the top 10, with Norway, Switzerland and Italy being the major consumers (Anon, 1995). The USA and Japan take 11th and 12th places, respectively.

## NEED FOR POSTHARVEST TECHNOLOGY
······

Fruit, vegetables and ornamentals are ideally harvested when eating or visual quality is at an optimum. However, since they are living biological systems they will deteriorate after harvest. The rate of deterioration varies greatly between individual produce depending on their overall rate of metabolism, but for many it can be rapid. For simple marketing chains, where produce is transferred from farm to end user in a short period, the rate of postharvest deterioration is of little consequence.

## TABLE 1.4 TOP 10 CUT FLOWERS AND POT PLANTS IN THE DUTCH AUCTION SYSTEM IN 1994

| RANK | CUT FLOWER | POT PLANT |
|------|------------|-----------|
| 1 | Rose | Ficus |
| 2 | Chrysanthemum | Dracaena |
| 3 | Tulip | Kalanchoe |
| 4 | Carnation | Begonia |
| 5 | Lily | Chrysanthemum |
| 6 | Gerbera | Azalea |
| 7 | Freesia | Hedera |
| 8 | Cymbidium | Saintpaulia |
| 9 | Alstroemeria | Spathiphyllum |
| 10 | Limonium | Yucca |

SOURCE Anon (1995) *Facts and figures about Dutch horticulture*, Flower Council of Holland, Leiden.

However, with the increasing remoteness of production areas from population centres in both developing and developed countries, the proliferation of large urban centres with complex marketing systems and the growth in international trading, the time from farm to market can be considerable. Adding to this time delay between farm and end user is the deliberate storage of certain produce to capture a better return by extending the marketing period into a time of shorter supply. Thus the modern marketing chain is providing increasing demands on produce and has created the need for postharvest techniques that allow quality to be retained over an increasingly longer period.

Extending the postharvest life of horticultural produce requires knowledge of all the factors that can lead to loss of quality or the generation of unsaleable material, as well as the use of this knowledge to develop affordable technologies that minimise the rate of deterioration. This field of scientific endeavour is now simply known as "postharvest". The increased attention given to postharvest horticulture in recent years has come from the realisation that faulty handling practices after harvesting can cause large losses of produce that required large inputs of labour, materials and capital to grow. Informed opinion now suggests that increased emphasis should be placed on conservation after harvest, rather than endeavouring to further boost crop production, as this would appear to offer a better return for the available resources of labour, energy and capital.

The actual causes of postharvest loss are many but can be classified into two main categories. The first of these is physical loss. This can arise from structural damage or microbial wastage, which can leave produce tissue degraded to a stage where it is not acceptable for presentation, fresh consumption or processing. Physical loss can also arise from the evaporation of intercellular water, which leads to a direct loss in weight. The resulting economic loss is primarily due to the reduced weight of produce that remains available for marketing, but can also be a reduced return obtained from a whole batch of commodity rejected through wastage of a small proportion of items in that batch.

Loss of quality is the second cause of postharvest loss, and this can be due to physiological and compositional changes that alter the appearance, taste or texture and make produce less desirable aesthetically to end users. The changes may arise from normal metabolism of produce or abnormal events arising from the postharvest environment. Economic loss is incurred in having to market such produce at a reduced price. In many markets there is no demand for second class produce even at a reduced price, which leads to a total economic loss even though it may still be edible.

In tropical regions, which include a large proportion of the developing countries, these losses can assume considerable economic and social importance. In developed regions such as North America, Europe and Australia/New Zealand, postharvest deterioration of fresh produce is often just as serious although often through different causes. As the value of fresh produce may increase many times from the farm to the retailer, the economic consequences of deterioration at any point along the chain are serious. When farms are located near towns and cities, faulty handling practices are often less important because the produce is usually consumed before serious wastage can occur. Even in the tropical regions, however, production of some staple commodities is seasonal, and there is a need to store produce to meet requirements during the "off" season. In industrialised countries and in countries that encompass a wide range of climatic regions, fresh fruit and vegetables are frequently grown at locations remote from the major centres of population. Thousands of tonnes of produce are now transported daily over long distances both within countries and internationally. Fresh fruit and vegetables are important items of commerce, and there is a huge investment of resources in transport, storage and marketing facilities designed to maintain a continuous supply of these perishable commodities. Postharvest technology aims to protect that investment.

# EXTENT OF POSTHARVEST LOSSES

The magnitude of losses of horticultural produce during postharvest and marketing operations are widely acknowledged to be considerable, although few studies have accurately quantified these losses. Part of the difficulty in quantifying postharvest losses is identifying the actual steps in the postharvest chain where the loss was induced. It is not uncommon for a physical or metabolic stress to be imposed on produce, but the visual deleterious action may not be evident until later in the marketing chain. For example, exposure to excess field heat after harvest can advance general senescence but visible symptoms, such as loss of green colour, may not occur for days or weeks. Also, the visible cause of loss may not be the actual cause; for example, chilling injury of tomatoes is induced by prolonged storage at sub-optimal temperatures, but visual symptoms are usually mould growth on the damaged tissues and not the chilling injury itself.

Any attempt to improve the postharvest handling of horticultural produce should be preceded by a quantitative estimate of postharvest losses. This will determine the economic and social imperatives for action and allow any intervention strategy to be assessed for cost/benefit. There is, however, little authoritative published data on postharvest losses. The most quoted document on fruit and vegetable losses was published by the US National Academy of Sciences in 1978, which reviewed the literature and expert opinion to that time. Values quoted range from 5–100 per cent loss of particular crops. While the findings may be the best estimate of losses on a global scale, they are virtually meaningless if applied to specific circumstances. A commonly repeated comment is that national postharvest losses are in the range of 20–40 per cent, usually made without any supporting quantitative data. A quantitative study organised by the ASEAN Food Handling Bureau (AFHB) for the 3 day transport of cabbages between Sumatra, Indonesia and Singapore found a loss of about 30 per cent (Lee et al., 1981); the difficulty of conducting such trials was emphasised by the need for the researchers to manually dissect 1.5 tonnes of cabbage in a 3 hour period after shipment. A quantitative study of losses at different points during marketing in Taiwan showed a range of 4–30 per cent, but different produce had differing susceptibility during transport, wholesale and retail operations (Table 1.5).

The Inter-American Institute for Cooperation on Agriculture (IICA), through La Gra (1990) and colleagues, has worked for many years on developing assessments of postharvest systems that are thorough but minimise the need for large scale quantitative measurements. An

TABLE 1.5  POSTHARVEST LOSSES IN SELECTED FRUIT
AND VEGETABLES DURING MARKETING IN TAIWAN

| | | % Loss | | |
| --- | --- | --- | --- | --- |
| COMMODITY | TRANSPORT | WHOLESALE | RETAIL | TOTAL |
| Chinese cabbage | 4 | 23 | 5 | 30 |
| Turnip | 2 | 10 | 4 | 15 |
| Green bean | 3 | 1 | 0 | 4 |
| Tomato | 1 | 5 | 2 | 8 |
| Watermelon | 11 | 1 | 0 | 12 |
| Muskmelon | 2 | 5 | 9 | 16 |
| Papaya | 2 | 7 | 14 | 21 |
| Carambola | 2 | 6 | 7 | 15 |
| Apple | 2 | 1 | 3 | 6 |
| Banana | 0 | 3 | 7 | 9 |

SOURCE  Liu, M.S. and P.C. Ma (1983) *Postharvest problems of vegetables and fruits in the tropics and subtropics*, 10th Anniversary Monograph Series, Asian Vegetable Research and Development Center, Shanhua, Taiwan.

inter-institutional collaboration between IICA, AFHB and the Postharvest Institute for Perishables (PIP) has resulted in the generation of a practical manual that details a systematic approach to identifying postharvest problems within any horticultural situation. Application of the methodology requires an interdisciplinary or team approach as knowledge is required of the preproduction, production, harvest, postharvest and marketing operations that make up any commodity handling system. It is claimed that a full appraisal can be conducted in 2–4 weeks and that will identify priority problems and alternate developments that can be ordered into priority solutions in a development strategy and time frame.

# POSTHARVEST TECHNOLOGY

The ultimate role of postharvest technology is to devise methods by which deterioration of produce is restricted as much as possible during the period between harvest and end use. This requires a thorough understanding of the structure, composition, biochemistry and physiology of horticultural produce as postharvest technologies will be mainly concerned with slowing down the rate of produce metabolism without inducing abnormal events. While there is a common underlying structure and metabolism, different types of produce vary in their response to specific postharvest situations. Appropriate postharvest technologies must be developed to cope with these differences. The variation in response can also be important between cultivars of the same produce and often between stages of maturity, growing areas or seasons.

The principal weapon in the postharvest armoury relates to control of the storage environment or handling conditions. Control over temperature is the most significant environmental factor as the rate of postharvest deterioration from all causes is affected by temperature. The response of produce to temperature is, however, not uniform over the normal temperature range of fresh horticultural produce, which is encompassed by the freezing point of plant tissues (about $-2°C$ to $0°C$) and an upper limit of about $40°C$ when plant tissues collapse. The deleterious effects of extreme temperatures, however, follow a time/temperature relationship, where produce can withstand abnormally high or low temperatures for a short period. The phenomenon is beneficially used during heat disinfestation treatments, whereby insect pests are killed following brief exposure of produce to temperatures above $45°C$ but produce suffer no adverse effect. The rate of normal metabolism decreases with decreasing temperature, but effects on produce quality are more pronounced at the lower end of the temperature scale; a change of $2°C$ at $20°C$ has much less effect on metabolism than the same temperature change at $3°C$. It would seem that the ideal postharvest temperature is just above the freezing point where metabolism is slowest. However, the onset of abnormal metabolism at reduced temperatures becomes a limiting factor for many produce. The beneficial effect of reduced temperature on microbial growth is an important consideration in postharvest systems.

Other important environmental conditions are the concentration of certain gases and water vapour in the atmosphere around produce. Maintenance of a high relative humidity in the atmosphere is necessary to minimise water loss, a key quality factor since wilted or shrivelled produce has greatly reduced market value. The use of modified and controlled atmospheres utilising elevated carbon dioxide and reduced oxygen levels from the normal atmosphere concentrations of 0 per cent and 21 per cent, respectively, has been known for many years to beneficially affect produce metabolism, but the problems of adequately containing gas levels within the beneficial range has restricted its use to pome fruit and a few other commodities. The recent development of plastic films with variable gas permeability and other atmosphere control features has reignited interest in modified atmosphere storage. The presence of ethylene in the atmosphere has been of concern in the postharvest handling of ornamentals and unripe climacteric fruit for many years because it promotes abscission, ripening and senescence, but its presence around non-climacteric fruit and vegetables is now also of renewed interest.

The major abnormal postharvest events are physiological disorders arising from adverse postharvest and preharvest environmental conditions or mineral imbalances during growth, and microbial decay from a range of bacteria and moulds that can infect produce either preharvest or postharvest. Apart from ensuring that produce is not exposed to the causative factors, control measures in the past have tended to focus on synthetic chemicals, but with current consumer concerns the thrust is now to the use of natural compounds or physical treatments. Postharvest insect infestation tends not to be a serious problem except where the insect is subject to quarantine restrictions, when it becomes a major technical and international and regional trade issue.

Apart from generating information on the effects of environmental conditions on particular produce, postharvest research must develop technology that is user friendly and cost effective to enable the benefits of scientific knowledge to have commercial value. This can be in the form of technology adaptation (such as the development of the refrigerated container, which created a mobile cool storage chamber) or the application of technical information to design new technology (such as the use of vapour-heat treatments for insect disinfestation). Packaging design can be considered as technology development since an essential requirement of packages is protection of the contents from physical damage and contamination while meeting other marketing criteria.

The need to be responsive to market needs has forced marked changes in quality assessment. This was originally limited to grading operations in the packing house for size or weight, removing defects and ensuring that containers are labelled correctly, but this is now a total quality operation to ensure that all market specifications are met and that the enterprise operates in the most efficient manner. Quality is thus linked with profitability and its successful implementation requires a complete understanding of all factors that affect produce and the market environment in which produce is traded. Thus, quality management starts in the field and continues until produce reaches the end user. Training of staff is now an integral part of quality management as individuals or work teams are empowered to take responsibility for ensuring that predetermined quality criteria are met.

Two recent innovations are having increasing impacts on postharvest technology. The first is the use of molecular biology to generate cultivars that overcome specific postharvest problems. The development of tomatoes that remain firm over an extended period due to inhibition of the polygalcturonase enzyme system and carnations with extended shelf life due to diminished production of ethylene are two such innovations that add to the substantial gains made by conventional plant breeding. Second, minimal processing of fruit and vegetables generates products that are cut or lightly processed in order to make the original commodity more easily used by consumers. This practice has been experiencing rapid growth, but invariably increases metabolism and renders produce more susceptible to microbial attack and environmental conditions.

# FURTHER READING
......

Anon (1995) *Facts and figures about Dutch horticulture*, Flower Council of Holland, Leiden.

Behe, B.K. (1993) Floral marketing and consumer research, *HortScience*, 28: 11–14.

Coursey, D.G. (1983) Postharvest losses in perishable foods of the developing world, in M. Lieberman (ed.) *Post-harvest physiology and crop preser-vation*, Plenum, New York.

Food and Agricultural Organization of the United Nations (1977) *An analysis of an FAO survey of post-harvest food losses in developing countries*, Rome.

Food and Agricultural Organization of the United Nations (1994) *FAO yearbook, trade, 1993*, vol. 47, Rome.

Food and Agricultural Organization of the United Nations (1995) *FAO yearbook, production, 1994*, vol. 48, Rome.

Kader, A.A., R.F. Kasmire, F.G. Mitchell, M.S. Reid, N.F. Sommer and J.F. Thompson (1992) *Postharvest technology of horticultural crops* (2nd ed.), University of California, Berkeley, CA. Special Publication 3311.

La Gra, J. (1990) *A commodity systems assessment for problem and project identification*, Postharvest Institute for Perishables, Moscow.

Lee, S.K., P.C. Leong, M. Soedibyo, Santoso and R.B.H. Wills (1981) Postharvest handling of cabbage from North Sumatra to Singapore, *Singapore J. Primary Production*, 9: 54–9.

National Academy of Sciences (1978) *Postharvest food losses in developing countries*, Washington, DC.

Shewfelt, R.L. and S.E. Prussia (eds) (1993) *Postharvest handling: A systems approach*, Academic Press, San Diego, CA.

Stephenson, R.A. (1989) *The Queensland Institute for Ornamental Horticulture: A conceptual study*, Queensland Department of Primary Industries Report, Brisbane. 93pp.

# 2
# STRUCTURE AND COMPOSITION

## STRUCTURE
......

The fleshy fruits of commerce comprise various combinations of tissues that may include an expanded ovary, the seed, and other plant parts such as the receptacle (e.g. apple, strawberry, cashew apple), bract and peduncle (e.g. pineapple). This combination of tissues gives us an organ defined by the *Shorter Oxford Dictionary* as 'the edible product of a plant or tree, consisting of the seed and its envelope, especially the latter when juicy and pulpy'. A consumer definition of fruit would be 'plant products with aromatic flavours, which are either naturally sweet or normally sweetened before eating'; they are essentially dessert foods. However, common usage has led to the consumption of immature fruit (e.g. zucchini, cucumber, beans) and even ripe fruit (e.g. avocado, tomato, capsicum, egg plant) as vegetables. These products, which have been referred to as fruit vegetables, may be consumed cooked or raw and are eaten either alone, in the form of a salad, or accompanying meat or fish dishes. However, common fruit are derived basically from an ovary and surrounding or associated tissues (Figure 2.1). Most of the exaggerated developments of certain parts of the fruit structure arose naturally but have been accentuated by breeding and selection to maximise the desirable features of each fruit and minimise the superfluous features. Naturally seedless cultivars of some fruits (e.g. banana, grape, navel orange) and others induced by breeding (e.g. watermelon) or management (e.g. persimmon) illustrate extreme development.

The vegetables do not represent any specific botanical grouping, and exhibit a wide variety of plant structures. They can, however, be grouped into three main categories: seeds and pods; bulbs, roots and tubers; flowers, buds, stems and leaves. In many instances, the structure giving rise to the particular vegetable has been highly modified compared with that structure on the 'ideal' plant. The derivation of some vegetables is shown in Figure 2.2. The plant part that gives rise to the vegetable will be readily apparent when most vegetables are visually examined. Some are a little more difficult to categorise, especially the tuberous organs developed underground. The potato, for instance, is a modified stem structure, but other underground storage organs, such as the sweet potato, are simply swollen roots.

The structural origins of fruit and vegetables have a major bearing on the recommendations for their postharvest preservation. Generally,

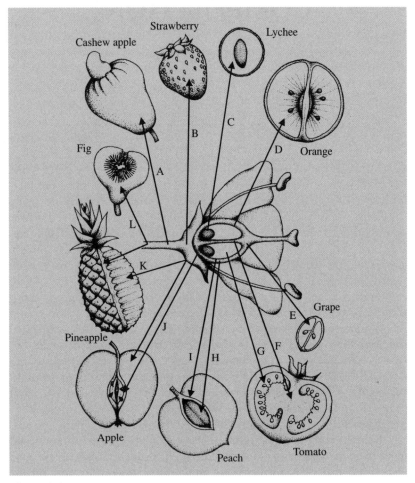

**Figure 2.1**
Derivation of some fruits from plant tissue. The letters indicate the tissues that comprise a significant portion of the fruit illustrated as follows: (A) pedicel, cashew apple; (B) receptacle, strawberry; (C) aril, lychee; (D) endodermal intralocular tissue, orange; (E) pericarp, grape; (F) septum, tomato: (G) placental intralocular tissue, tomato; (H) mesocarp, peach; (I) endocarp, peach; (J) carpels, apple; (K) accessory tissue, apple and pineapple; (L) peduncle, pineapple and fig.

above-ground structures develop natural wax coatings as they mature that reduce transpiration, whereas roots do not develop such coatings and therefore should be stored at high relative humidity (RH) to minimise water loss. Tuberous vegetables are equipped with a special capability to heal wounds caused by natural insect attack. This property is useful for minimising damage inflicted on tubers during harvesting.

Species of flowering plants used commercially as cut flowers have been selected for their visual appeal. At both the practical and botanical levels, cut flowers are variations of inflorescences. Although there is a huge range of variation in flower structure, the basic structure of an inflorescence is stem, including pedicels and peduncles, bracts and flowers. Figure 2.3 illustrates the range of variation in cut flower types. The range includes solitary to multiple inflorescences, where all inflorescences develop at about the same rate or where there is a gradation from mature to juvenile flowers as the inflorescence develops. It is important to recognise the variations in the structure of inflorescences because they have a major bearing on postharvest handling strategies. Generally, inflorescences have a low carbohydrate reserve compared with most fruits, although it can be similar to many leafy vegetables. In many cases improved vase or storage life can be achieved by providing the cut flower with sugar through an absorbed solution. Because of their enormous surface area compared with their mass, cut flowers transpire at much higher rates than most fruit.

# CELLULAR COMPONENTS

The cells of fruit and vegetables are typical plant cells, the principal components of which are shown in Figure 2.4. A brief outline of the essential features or functions of these components will be given; more detailed explanations can be found in specialised texts.

Plant cells are bounded by a more or less rigid cell wall composed of cellulose fibres, and other polymers such as pectic substances, hemicelluloses, lignins and proteins. A layer of pectic substances forms the middle lamella and acts to bind adjacent cells together. Adjacent cells often have small communication channels, called plasmadesmata, linking their cytoplasmic masses. The cell wall is permeable to water and solutes. The main functions of the wall are to:

1.  contain the cell contents by supporting the outer cell membrane, the plasmalemma, against the hydrostatic pressures of the cell contents, which would otherwise burst the membrane; and

2.  give structural support to the cell and the plant tissues.

Within the plasmalemma, the cell contents comprise the cytoplasm and usually one or more vacuoles. The latter are fluid reservoirs containing various solutes — such as sugars, amino and organic acids, and salts—and are surrounded by a semipermeable membrane, the

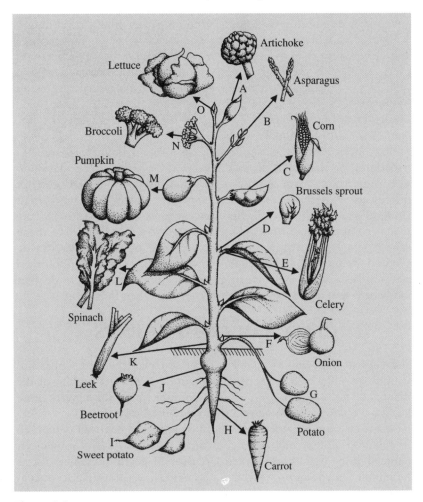

**Figure 2.2**

Derivation of some vegetables from plant tissue. The letters indicate the principal origins of representative vegetables as follows: (A) flower bud, artichoke; (B) stem sprout, asparagus; (C) seeds, corn; (D) axillary bud, brussels sprout; (E) petiole, celery; (F) bulb (underground bud), onion; (G) stem tuber, potato; (H) swollen root, carrot; (I) swollen root tuber, sweet potato; (J) swollen hypocotyl, beetroot; (K) swollen leaf base, leek; (L) leaf blade, spinach; (M) fruit, pumpkin; (N) swollen inflorescence, broccoli; (O) main bud, lettuce.

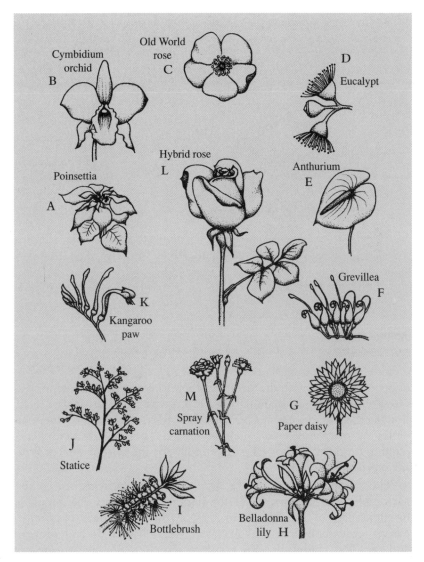

**Figure 2.3**

Examples of the variations in the structure of flowers: (A) bract, poinsettia; (B) modifications and fusions, cymbidium orchid in which the labellum is a median modified petal and the column is comprised of fused stamens and pistils; (C) complete, single whorl of petals, old world rose; (D) prominent feature (stamens), eucalypt; (E) spadix plus spathe, anthurium; (F) raceme, grevillea; (G) head, paper daisy; (H) umbel, belladonna lily; (I) spike, bottlebrush; (J) panicle, statice; (K) cyme, kangaroo paw; (L) solitary, hybrid tea rose; (M) corymb, spray carnation.

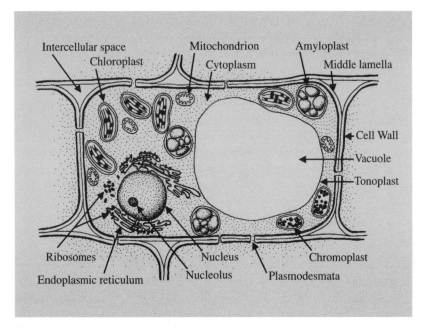

**Figure 2.4**
Diagrammatic representation of a plant cell and its constituent organelles.

tonoplast. Together with the semipermeable plasmalemma, the tonoplast is responsible for maintaining the hydrostatic pressure of the cell, allowing the passage of water but selectively restricting the movement of solutes or macromolecules, such as proteins and nucleic acids. The resulting turgidity of the cell is responsible for the crisp nature of fruit and vegetables.

The cytoplasm comprises a fluid matrix of proteins and other macromolecules and various solutes. Important processes that occur in this fluid part of the cytoplasm include the breakdown of storage reserves of carbohydrate by glycolysis (see Chapter 3) and protein synthesis. The cytoplasm also contains several important organelles, which are membrane-bound bodies with specialised functions.

1. The nucleus is the largest organelle. It is the control centre of the cell, containing the genetic information in the form of DNA (deoxyribonucleic acid). It is bounded by a porous membrane that has distinct holes when viewed under the electron microscope. These permit the movement of mRNA (messenger ribonucleic acid) — the transcription product of the genetic code of DNA —

into the cytoplasm, where mRNA is translated into proteins on the ribosomes of the protein synthesising system (see below).

2.  Mitochondria contain the respiratory enzymes of the tricarboxylic acid (TCA) cycle (see Chapter 3) and the respiratory electron transport system which synthesise adenosine triphosphate. Mitochondria utilise the products of glycolysis for energy production. Thus they form the energy powerhouse of the cell.

3.  Chloroplasts, found in green cells, are the photosynthetic apparatus of the cell. They contain the green pigment chlorophyll and the photochemical apparatus for converting solar energy (light) into chemical energy. As well, they have the enzymes necessary for fixing atmospheric carbon dioxide to synthesise sugars and other carbon compounds.

4.  Chromoplasts develop mainly from mature chloroplasts when the chlorophyll is degraded. They contain carotenoids, which are the yellow-red pigments in many fruits.

5.  Amyloplasts are the sites of starch grain development, although starch grains are also found in chloroplasts. Collectively chloroplasts, chromoplasts and amyloplasts are known as plastids.

6.  The Golgi complex is a series of plate-like vesicles that bud off smaller vesicles. These are probably of importance in cell wall synthesis and the secretion of enzymes from the cell.

7.  The endoplasmic reticulum is a network of tubules within the cytoplasm. Some evidence suggests it may act as a transport system in the cytoplasm. What is more clear is that ribosomes — the sites of protein synthesis— are often attached to the endoplasmic reticulum. Other ribosomes are found free in the cytoplasm. The ribosomes contain ribonucleic acids and proteins.

# CHEMICAL COMPOSITION AND NUTRITIONAL VALUE OF FRUIT AND VEGETABLES[1]

·······

## *Water*

Most fruit and vegetables contain more than 80 g of water/100 g of produce, with some, such as cucumber, lettuce, marrow and melons, containing about 95 g of water/100 g of produce. The starch tubers and

1 All values are expressed on a fresh weight basis.

seeds, such as yam, cassava and corn, contain less water, but even they usually comprise more than 50 g/100 g of produce. Quite large variations in water content can occur within a species since the water content of individual cells varies considerably. The actual water content depends on the availability of water to the tissue at the time of harvest, so that the water content of produce will vary during the day if there are diurnal fluctuations in temperature and relative humidity. For most produce it is desirable to harvest when the maximum possible water content is present, as this results in a crisp texture. Hence the time of harvest can be an important consideration, particularly with leafy vegetables, which exhibit large and rapid variations in water content in response to changes in their environment.

## *Carbohydrates*

Carbohydrates are generally the most abundant constituents after water. They can be present across a wide molecular weight range from simple sugars to complex polymers, which may comprise many hundreds of sugar monomeric units. Carbohydrates can account for 2–40 g/100 g of produce tissue, with low levels being found in some cucurbits, such as cucumber, and high levels in vegetables that accumulate starch, such as cassava.

The main sugars present in fruit and vegetables are sucrose, glucose and fructose, with the predominant sugar varying in different produce

### TABLE 2.1  SUGAR CONTENT (G/100 G FRESH WEIGHT) OF SOME RIPE FRUITS AND VEGETABLES

| PRODUCE | GLUCOSE | FRUCTOSE | SUCROSE |
|---------|---------|----------|---------|
| Apple | 3 | 6 | 2 |
| Banana | 4 | 4 | 10 |
| Beetroot | < 1 | < 1 | 8 |
| Capsicum | 2 | 2 | 0 |
| Cherry | 6 | 4 | 0 |
| Grape | 8 | 8 | 0 |
| Onion | 2 | 2 | 1 |
| Orange | 2 | 2 | 4 |
| Pea | < 1 | < 1 | 4 |
| Peach | 1 | 1 | 5 |
| Pear | 2 | 7 | 1 |
| Pineapple | 1 | 2 | 5 |
| Tomato | 1 | 1 | 0 |

Note: Zero indicates <0.1 g/100 g present.

(Table 2.1). Glucose and fructose occur in all produce and are often present at a similar level, while sucrose is only present in about two-thirds of produce. Produce with the highest sugar levels (Table 2.2) are mainly tropical and subtropical fruit, with grape the only temperate fruit listed and no vegetables listed. Beetroot contains the highest sugar content among the vegetables at about 8 g/100 g, with sucrose the only sugar present. Much of the sensory appeal of fruit is due to the sugar content as sugars produce a sweet taste that is considered to be one of the universal innate human taste preferences

Humans can digest and utilise sugars and starch as energy sources, and hence vegetables with a high starch content are important contributors to the daily energy requirement of people in many societies. Produce such as cassava and yam commonly contain more than 20 g/100 g as starch, with other starchy produce containing more than 10 g/100 g as starch. Starch from plantain, cassava, yam, taro, sweet potato and potato provides the bulk of energy in simple diets of subsistence groups in some developing countries. In these diets, over-dependence on the starchy vegetables is undesirable as they cannot supply enough of the other essential nutrients. However, consumers in developed countries are being encouraged to eat more starch — or complex carbohydrates as it is now called — although in these countries cereals rather than fruit and vegetables are the major source of dietary starch.

A substantial proportion of carbohydrates is present as dietary fibre, which is not digested in the human upper intestinal system but is either

## TABLE 2.2 FRUIT AND VEGETABLES WITH THE HIGHEST SUGAR LEVELS (G/100 G)

| PRODUCE | TOTAL SUGARS | GLUCOSE | FRUCTOSE | SUCROSE |
|---|---|---|---|---|
| Banana | 17 | 4 | 4 | 10 |
| Jackfruit | 16 | 4 | 4 | 8 |
| Litchi | 16 | 8 | 8 | 1 |
| Persimmon | 16 | 8 | 8 | 0 |
| Rambutan | 16 | 3 | 3 | 10 |
| Grape | 15 | 8 | 8 | 0 |
| Custard apple | 15 | 5 | 6 | 4 |
| Pomegranate | 14 | 8 | 6 | 0 |
| Carambola | 12 | 1 | 3 | 8 |
| Mango | 12 | 1 | 3 | 8 |

Note: Some differences arise between values given for total sugars and the total of individual sugars due to rounding of data given in R.B.H. Wills (1987) Composition of Australian fresh fruit and vegetables, *Food Technol. Aust.* 39: 523–6.

metabolised in the lower intestinal system or passes from the body in the faeces. Cellulose, pectic substances and hemicelluloses are the main carbohydrate polymers that constitute fibre. Lignin, a complex polymer of aromatic compounds linked by propyl units, is also a major component of dietary fibre. Dietary fibre is not digested as humans are not capable of secreting the enzymes necessary to break down the polymers to the basic monomeric units that can be absorbed by the intestinal tract. Starch and cellulose have the same chemical composition as they are synthesised from D-glucose units, but the bonding between the monomers differs. Starch comprises $\alpha$-1,4 linkages, which are hydrolysed by a range of amylase enzymes secreted by humans. Cellulose is formed with ß-1,4 linkages, but cellulase enzymes are not produced by humans. Similarly humans lack the enzymes necessary to degrade the pectins and hemicelluloses to galacturonic acid units, and xylose and other pentose constituents, respectively. Dietary fibre was once considered to be an unnecessary component in the diet, although it was known to relieve constipation. However, increased consumption of dietary fibre is now actively promoted by health agencies to prevent or alleviate a raft of diseases that mainly afflict Western society: the so-called diseases of affluence (Table 2.3).

## Protein

Fresh fruit and vegetables generally contain about l g protein/100 g in fruit and about 2 g/100 g in most vegetables, with the most abundant protein sources being the *Brassica* vegetables, which contain 3–5 g protein/100 g, and the legumes, which contain about 5 g protein/100 g. The protein is mostly functional, such as in the form of enzymes, rather than acting as a storage pool, as in grains and nuts. Their relatively low level means that fresh fruit and vegetables are not an important source of protein in the diet.

## Lipids

Lipids comprise less than 1 g/100 g of most fruit and vegetables and are

### TABLE 2.3   DISEASES CLAIMED TO BE DUE TO LACK OF DIETARY FIBRE

| | |
|---|---|
| Appendicitis | Haemorrhoids |
| Colon cancer | Hiatus hernia |
| Constipation | Ischaemic heart disease |
| Deep vein thrombosis | Obesity |
| Diabetes | Rectal tumours |
| Diverticulosis | Varicose veins |
| Gallstones | |

associated with protective cuticle layers on the surface of the produce and with cell membranes. The avocado and olive (used as a fresh vegetable) are exceptions, having respectively about 20 and 15 g lipid /100 g present as oil droplets in the cells. The generally low lipid content is seen as a positive factor in combating the rise of heart disease in the community, and increased consumption of fruit and vegetables is extensively promoted by health authorities. Even produce with a relatively high lipid content, such as avocados, are considered benign at worst as the dominant fatty acids are monounsaturated, which have been shown not to increase the incidence of heart disease.

## Organic acids

Most fruit and vegetables contain organic acids at levels in excess of what is required for the operation of the TCA cycle and other metabolic pathways. The excess is generally stored in the vacuole away from other cellular components. Lemon, lime, passionfruit and black currant often contain more than 3 g organic acids/100 g. The dominant acids in produce are usually citric and malic acid, and some examples are given in Table 2.4. Other organic acids that are dominant in certain commodities are tartaric acid in grapes, oxalic acid in spinach and isocitric acid in blackberries. Apart from their biochemical importance, organic acids contribute greatly to taste, particularly of fruit, with a balance of sugar and acid giving rise to the desirable taste of specific produce.

## Vitamins and minerals

Vitamin C (ascorbic acid) is only a minor constituent of fruit and vegetables but is of major importance in human nutrition for the prevention of the disease scurvy. Virtually all human dietary vitamin C (approximately 90 per cent) is obtained from fruit and vegetables. The daily requirement for vitamin C is about 50 mg, and many commodities

### TABLE 2.4  SOME FRUIT & VEGETABLES IN WHICH CITRIC & MALIC ACIDS ARE THE MAJOR ACIDS PRESENT

| CITRIC | | MALIC | |
| --- | --- | --- | --- |
| Berries | Beetroot | Apple | Broccoli |
| Citrus | Leafy vegetables | Banana | Carrot |
| Guava | Legumes | Cherry | Celery |
| Pear | Potato | Melon | Lettuce |
| Pineapple | Plum | Onion | |
| Tomato | | | |

contain this amount of vitamin C in less than 100 g of tissue.

Fruit and vegetables may also be important nutritional sources of vitamin A and folic acid, commonly supplying about 40 per cent of daily requirements. Vitamin A is required by the body to maintain the structure of the eye. A prolonged deficiency of vitamin A can eventually lead to blindness. The active vitamin A compound, retinol, is not present in produce, but some carotenoids such as ß-carotene can be converted to retinol by humans. Only about 10 per cent of the carotenoids known to be in fruit and vegetables are precursors of vitamin A. All other carotenoids such as lycopene, the main pigment in tomato, have no vitamin A activity. Recent epidemiological and laboratory studies have implicated a role for carotenoids in the prevention of various forms of cancer, although there is now conflicting evidence on this. Initial interest was in ß-carotene, but the oxygenated carotenoids, the xanthophylls, particularly lutein, are becoming regarded as being of greater importance.

Folic acid is involved in RNA synthesis, and a deficiency will result in anaemia. Folic acid deficiency during early pregnancy has been associated with foetal spina bifida, and various countries have moved to fortify certain foods with folic acid to minimise this risk. Green leafy vegetables are good sources of folic acid, with the intensity of green colour acting as a good guide to the folic acid content. Table 2.5 shows the range of levels of vitamins C and A and folic acid in some fruit and vegetables. Maintenance of these vitamins during handling and storage should be a major concern, particularly when the produce will be consumed by people on marginally sufficient diets.

The major mineral in fruit and vegetables is potassium, which is present at more than 200 mg/100 g in most produce. The highest levels are in green leafy vegetables, with parsley containing about 1200 mg/100 g, but about 20 vegetables contain 400-600 mg/100 g. Health authorities in many countries are urging increased consumption of potassium to counter the effects of sodium in the diet, and fruit and vegetables are the richest natural food source of potassium.

Many other vitamins and essential minerals are present in fruit and vegetables, but their contribution to total dietary requirements is generally of minor importance. Iron and calcium may be present at nutritionally significant levels, although often in a form that is unavailable for absorption by humans. For example, most of the calcium in spinach is present as calcium oxalate, which is only poorly absorbed.

The nutritional value of various fruits and vegetables depends not only on the concentration of nutrients in the produce but also on the

## TABLE 2.5 LEVELS OF VITAMIN C, VITAMIN A AND FOLIC ACID IN SOME FRUITS AND VEGETABLES

| COMMODITY | VITAMIN C (MG/100 G) | COMMODITY | VITAMIN A B-CAROTENE EQUIVALENT (MG/100) | COMMODITY | FOLIC ACID (μG/100 G) |
|---|---|---|---|---|---|
| Black currant, guava | 200 | Carrot | 10.0 | Spinach | 80 |
| Chilli | 150 | Sweet potato (red) | 6.8 | Broccoli | 50 |
| Broccoli, Brussels sprout | 100 | Parsley | 4.4 | Brussels sprout, pulses | 30 |
| | | Spinach | 2.3 | | |
| Papaya | 80 | Mango | 2.4 | Cabbage, lettuce | 20 |
| Kiwi fruit | 70 | Red chilli | 1.8 | | |
| Citrus, strawberry | 40 | Tomato | 0.3 | Banana | 10 |
| Cabbage, lettuce | 35 | Apricot | 0.1 | Most fruits | < 5 |
| Mango, carrot | 30 | Banana | 0.1 | | |
| Pineapple, banana, potato, tomato, cassava | 20 | Potato | 0.0 | | |
| Apple, peach | 10 | | | | |
| Beetroot, onion | 5 | | | | |

amount of such produce consumed in the diet. An attempt to equate these factors and show the relative concentration of 10 major vitamins and minerals in some fruits and vegetables and their importance in the typical US diet is shown in Table 2.6. Tomatoes and oranges are relatively low in concentration of nutrients but make the major contribution of all produce to US diets because of the large per capita consumption.

## *Volatiles*

All fruit and vegetables produce a range of small molecular weight compounds (molecular weight less than 250) that possess some volatility at ambient temperatures. These compounds are not important quantitatively (normally less than 10 mg/100 g are present), but they are important in producing the characteristic flavour and aroma of fruit and,

### TABLE 2.6 RELATIVE CONCENTRATION OF 10 VITAMINS AND MINERALS IN FRUIT & VEGETABLES & THE RELATIVE CONTRIBUTION OF VITAMINS & MINERALS THESE COMMODITIES MAKE TO THE US DIET

| NUTRIENT CONCENTRATION | | CONTRIBUTION OF NUTRIENTS TO DIET | |
|---|---|---|---|
| CROP | RANK | CROP | RANK |
| Broccoli | 1 | Tomato | 1 |
| Spinach | 2 | Orange | 2 |
| Brussels sprout | 3 | Potato | 3 |
| Lima bean | 4 | Lettuce | 4 |
| Pea | 5 | Sweet corn | 5 |
| Asparagus | 6 | Banana | 6 |
| Artichoke | 7 | Carrot | 7 |
| Cauliflower | 8 | Cabbage | 8 |
| Sweet potato | 9 | Onion | 9 |
| Carrot | 10 | Sweet potato | 10 |
| Sweet corn | 12 | Pea | 15 |
| Potato | 14 | Spinach | 18 |
| Cabbage | 15 | Broccoli | 21 |
| Tomato | 16 | Lima bean | 23 |
| Banana | 18 | Asparagus | 25 |
| Lettuce | 26 | Cauliflower | 30 |
| Onion | 31 | Brussels sprout | 34 |
| Orange | 33 | Artichoke | 36 |

SOURCE Adapted from C.M.Rick, (1978) The tomato, *Sci. Am.* 239(2): 66–76.

to a lesser extent, of vegetables. Aroma constituents of many ornamentals have found application as perfumes for human use or as background odours in manufactured products. Most fruits and vegetables and many ornamentals each contain in excess of 100 different volatiles, and the number of compounds identified in particular produce is continually increasing as the sensitivity of the analytical techniques for their identification improves. The compounds are mainly esters, alcohols, acids and carbonyl compounds (aldehydes and ketones). Many of these compounds, such as ethanol, are common to all fruit and vegetables while others are specific to an individual or species. Esters, for example, are common constituents of most ripe fruits while sulphur-containing volatiles are present in *Brassica* vegetables.

Definitive studies correlating consumer recognition of the produce with the volatile profile emanating from the produce have shown that only a small number of compounds are responsible for consumer recognition of that commodity. In most fruit and vegetables the characteristic aroma is due to the presence of one or two compounds.

## TABLE 2.7 DISTINCTIVE COMPONENTS OF THE AROMA OF SOME FRUITS AND VEGETABLES

| PRODUCT | | COMPOUND |
|---|---|---|
| Apple | — ripe | Ethyl 2-methylbutyrate |
| | — green | Hexanal, 2-hexenal |
| Banana | — green | 2-Hexenal |
| | — ripe | Eugenol |
| | — overripe | Isopentanol |
| Grapefruit | | Nootakatone |
| Lemon | | Citral |
| Orange | | Valencene |
| Raspberry | | 1-($\pi$-Hydroxyphenyl)-3-butanone |
| Cucumber | | 2,6-Nonadienal |
| Cabbage — raw | | Allyl isothiocyanate |
| — cooked | | Dimethyl disulphide |
| Mushroom | | 1-Octen-3-ol, lenthionine |
| Potato | | 2-Methoxy-3-ethyl pyrazine, 2.5-dimethyl pyrazine |
| Radish | | 4-Methylthio-*trans*-3-butenyl sothiocyanate |

SOURCE Adapted from D.K. Salunkhe, J.Y. Do (1977) Biogenesis of aroma constituents of fruits and vegetables, *CRC Crit. Rev. Food Sci. Nutr,* 8 161–90.

### TABLE 2.8 COMPOSITION OF SELECTED FRUIT & VEGETABLES

| PRODUCE | WATER | PROTEIN | FAT | SUGARS | STARCH | DIETARY FIBRE | VIT. C |
|---|---|---|---|---|---|---|---|
| Apple | 84.3 | 0.3 | 0.1 | 11.7 | 0.2 | 2.3 | 5 |
| Banana — ripe | 75.2 | 1.7 | 0.1 | 17.3 | 3.1 | 2.8 | 12 |
| Banana — unripe | 71.9 | 1.9 | 0.1 | 1.3 | 21.2 | 3.2 | 18 |
| Broccoli | 88.7 | 4.7 | 0.3 | 0.4 | 0 | 3.9 | 106 |
| Carrot | 88.4 | 0.7 | 0.1 | 5.4 | 0 | 3.6 | 6 |
| Lettuce | 95.8 | 0.9 | 0.1 | 0.4 | 0 | 1.7 | 4 |
| Mango | 82.8 | 1.0 | 0.2 | 12.1 | 0.5 | 1.6 | 28 |
| Parsley | 89.1 | 2.2 | 0.2 | 0.4 | 0 | 5.4 | 114 |
| Potato | 81.2 | 2.3 | 0.1 | 0.4 | 12.6 | 2.2 | 21 |
| Tomato | 94.7 | 1.0 | 0.1 | 1.9 | 0 | 1.6 | 18 |

Table 2.7 gives the key compounds claimed to be responsible for the characteristic aromas of some fruit and vegetables. Practically all the compounds mentioned in Table 2.7 are minor components of the aroma fraction. The olfactory senses are thus extremely sensitive. The threshold concentration, or minimum concentration, at which the odour of ethyl 2-methylbutyrate, the main characteristic odour of apple, can be detected organoleptically was found to be 0.001 $\mu$L/L; that is, an apple weighing 100 g is recognised if 0.01 $\mu$g of ethyl 2-methylbutyrate is present. For the characteristic odour to be desirable it must also be in the correct concentration. At different stages of maturation, different compounds become the dominant component of flavour. Thus a

| B-CAROTENE | POTASSIUM | CALCIUM | |
|---|---|---|---|
| 0 | 100 | 4 | **NOTE** |
| 0.1 | 350 | 5 | Units are g/100 g edible portion of |
| | | | macronutrients and mg/100 g edible |
| 0.2 | 320 | 5 | portion of vitamins and minerals. |
| | | | **SOURCE** |
| 0.4 | 360 | 31 | Adapted from R.B.H. Wills (1987) |
| 10.0 | 240 | 3 | Composition of Australian fresh fruit |
| 0.1 | 230 | 16 | and vegetables, *Food Technol. Aust.*, 39: |
| 2.4 | 250 | 7 | 523-26; and R.B.H. Wills, J.S.K. Lim |
| 4.4 | 1190 | 200 | and H. Greenfield (1984) Changes in |
| | | | chemical composition of 'Cavendish' |
| 0 | 430 | 3 | banana (*Musa acuminata*) during |
| 0.3 | 200 | 8 | ripening, *J. Food Biochem*, 8: 69-77. |

blindfolded subject would be able to detect the stage of development of a particular commodity by sniffing the aroma. Some examples of these changes are given in Table 2.7.

The composition of a small selection of fruit and vegetables is given in Table 2.8 to illustrate the range in values of individual components that may be encountered. Details of the chemical composition and nutritional value of fruit and vegetables are to be found in many publications. Much of the data from these sources are variable because of inherent differences between cultivars and because of the effects of maturity, season and locality. The values presented here should therefore be regarded only as a guide to the composition of fruit and vegetables.

# FURTHER READING

Clark, I.and H. Lee (1989) *Name that flower*, Melbourne University, Melbourne.

Gehhardt, S.E., R. Cutrifelli and R.H. Matthews, (1982) *Composition of foods: Fruits and fruit juices, raw, processed, prepared*, Agriculture Handbook, no. 8–9, US Department of Agriculture, Washington, DC.

Haytowitz, D.B. and R.H. Matthews (1984) *Composition of foods: Vegetables and vegetable products, raw, processed, prepared*, Agriculture Handbook, no. 8–11, US Department of Agriculture, Washington, DC.

Hedrick, U.P. (1972) *Sturtevant's edible plants of the world*, Dover Publication, New York.

Holland, B., A.A. Welch, I.D. Unwin, D.H. Buss, A.A. Paul and D.A.T. Southgate, (1991) *McCance and Widdowson's The composition of foods* (5th rev. ed.), Royal Society of Chemistry, Cambridge, UK.

Kays, S.J. (1991) *Postharvest physiology of perishable plant products*, Van Nostrand Reinhold, New York.

McCarthy, M.A. and R.H. Matthews (1984) *Composition of foods: Nut and seed products*, Agriculture Handbook, no. 8–12, US Department of Agriculture, Washington, DC.

Maarse, H. (ed) (1991) *Volatile compounds in foods and beverages*, Marcel Dekker, New York.

Passmore, R. and M.A. Eastwood (1986) *Davidson and Passmore's human nutrition and dietetics* (8th ed.), Churchill Livingstone, Edinburgh.

Seymour, G.B., E. Taylor and G.A. Tucker (1993) *Biochemistry of fruit ripening*, Chapman & Hall, London.

Wills, R.B.H. (1987) Composition of Australian fresh fruit and vegetables, *Food Technol. Aust.* 39: 523–26.

# 3
# PHYSIOLOGY AND BIOCHEMISTRY

An important yet basic fact regarding the postharvest handling of horticultural produce is that they are 'living' structures. One readily accepts that produce is a living, biological entity when it is attached to the growing parent plant in its agricultural environment, but even after harvest the produce is still living as it continues to perform most of the metabolic reactions and maintain the physiological systems that were present when it was attached to the plant.

An important feature of plants — and therefore of fruit, vegetables and ornamentals — is that they respire by taking up oxygen and giving off carbon dioxide and heat. They also transpire (i.e. lose) water. While attached to the plant, the losses due to respiration and transpiration are replaced from the flow of sap, which contains water, photosynthates (principally sucrose and amino acids) and minerals. Respiration and transpiration continue after harvest, and since the produce is now removed from its normal source of water, photosynthates and minerals, the produce is dependent entirely on its own food reserves and moisture content. Therefore, losses of respirable substrates and moisture are not replenished and deterioration commences. In other words, harvested fruit and vegetables and ornamentals are perishable.

This chapter will consider the postharvest behaviour of horticultural commodities with particular reference to the physiological and biochemical changes that occur in ripening fruits. For this discussion, some understanding of the physiological development of fruit and vegetables and ornamentals is necessary.

## PHYSIOLOGICAL DEVELOPMENT

The life of fruit and vegetables can be conveniently divided into three major physiological stages following germination. These are growth, maturation and senescence (Figure 3.1). However, clear distinction between the various stages is not easily made. Growth involves cell division and subsequent cell enlargement, which accounts for the final

**Figure 3.1**

Growth, respiration and ethylene production patterns of climacteric and non-climacteric plant organs.

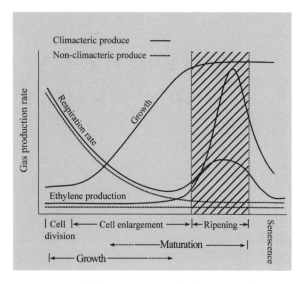

size of the produce. Maturation usually commences before growth ceases and includes different activities in different commodities. Growth and maturation are often collectively referred to as the development phase. Senescence is defined as the period when anabolic (synthetic) biochemical processes give way to catabolic (degradative) processes, leading to ageing and finally death of the tissue. Ripening — a term reserved for fruit — is generally considered to begin during the later stages of maturation and to be the first stage of senescence. The change from growth to senescence is relatively easy to delineate. Often the maturation phase is described as the time between these two stages, without any clear definition on a biochemical or physiological basis.

It is difficult to assign specific biochemical or physiological parameters to delineate the various stages because the parameters for different commodities are not identical in their nature or timing. Figure 3.2 shows the major changes in certain biochemical and physiological parameters in the climacteric tomato as it ripens from mature green to ripe. Similar changes are found in many non-climacteric plant tissues such as the pineapple.

Development and maturation of fruit are completed only when it is attached to the plant, but ripening and senescence may proceed on or off the plant. Fruit are generally harvested either when mature or when ripe, although some fruits that are consumed as vegetables may be harvested even before maturation has commenced, such as zucchini.

Similar terminology may be applied to the vegetables or

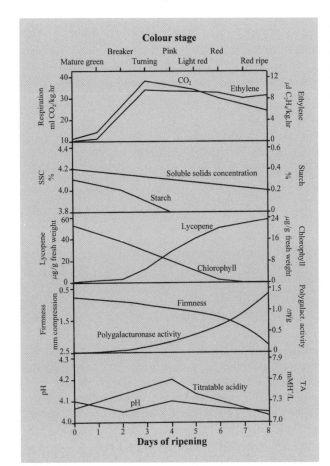

**Figure 3.2**
Physiochemical changes that occur during ripening of harvested tomatoes at 20°C.

ornamentals, or to any determinant organ, except that the ripening stage does not occur. As a consequence it is more difficult to delineate the change from maturation to senescence in vegetables and ornamentals. Vegetables and ornamentals are harvested over a wide range of physiological ages, that is, from a time well before the commencement of maturation through to the commencement of senescence (see Figure 10.3 on page 169).

# FRUIT RIPENING

·······

Ripening fruit undergoes many physicochemical changes after harvest that determine the quality of the fruit purchased by the consumer.

## TABLE 3.1 CHANGES THAT MAY OCCUR DURING THE RIPENING OF FLESHY FRUIT

Seed maturation
Colour changes
Abscission (detachment from parent plant)
Changes in respiration rate
Changes in rate of ethylene production
Changes in tissue permeability and cellular compartmentation
Softening: changes in composition of pectic substances
Changes in carbohydrate composition
Organic acid changes
Protein changes
Production of flavour volatiles
Development of wax on skin

SOURCE Adapted from H.K. Pratt (1975) *The role of ethylene in fruit ripening, Facteurs et régulation de la maturation des fruits*, Centre National de La Recherche Scientifique, Anatole, France. pp. 153–60.

Ripening is a dramatic event in the life of a fruit — it transforms a physiologically mature but inedible plant organ into a visually attractive olfactory and taste sensation. Ripening marks the completion of development of a fruit and the commencement of senescence, and it is normally an irreversible event. The following sections will discuss the general nature of fruit ripening, respiratory behaviour and the involvement of the gas ethylene $(C_2H_4)$ with these processes.

Ripening is the result of a complex of changes, many of them probably occurring independently of one another. A list of the major changes that together make up fruit ripening is given in Table 3.1. The time course of some of these changes is shown in Figure 3.3 for the banana, which is a climacteric fruit. The principal difference between the climacteric tomato and banana and the non-climacteric pineapple is the presence of the respiratory peak that is characteristic of climacteric fruits. A sharp increase in respiration is shown by the increase in the production of carbon dioxide or decrease in the internal oxygen concentration. Two of the changes listed — namely respiration and ethylene production — have gained priority in attempts to develop an explanation of the mechanism of fruit ripening. Further characterisation of other changes occurring in climacteric and non-climacteric fruit is given later in this chapter (see 'Chemical changes during maturation' on page 54).

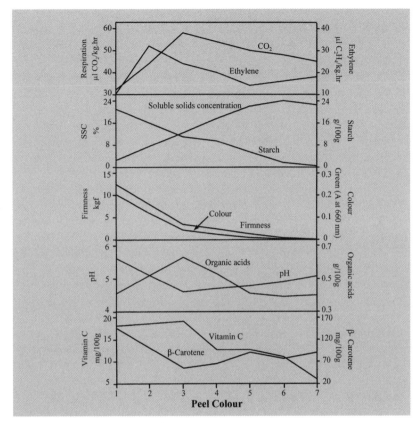

**Figure 3.3**

Physiochemical changes that occur during ripening of the Cavendish banana (variety Williams). The peel colour stages indicate the change from green (stage 1) to full yellow (stage 6), and finally to a stage when skin spotting occurs (stage 7). Williams bananas take about 8 days to progress from stage 1 to stage 7 at 20°C.

SOURCE Adapted from R.B.H. Wills, J.S.K. Lim and H. Greenfield (1984) Changes in chemical composition of "Cavendish" banana (*Musa acuminata*) during ripening, *J. Food Biochem.* 8: 69–77.

## Physiology of respiration

A major metabolic process taking place in harvested produce or in any living plant product is respiration. Respiration can be described as the oxidative breakdown of the more complex materials normally present in cells, such as starch, sugars and organic acids, into simpler molecules, such as carbon dioxide and water, with the concurrent production of energy and other molecules that can be used by the cell for synthetic reactions. Respiration can occur in the presence of oxygen (aerobic respiration) or in the absence of oxygen (anaerobic respiration, which is sometimes called fermentation).

The respiration rate of produce is an excellent indicator of metabolic activity of the tissue and thus is a useful guide to the potential storage life of the produce. If the respiration rate of a fruit or vegetable is measured — as either oxygen consumed or carbon dioxide evolved — during the course of its development, maturation, ripening and senescent periods, a characteristic respiratory pattern is obtained. Respiration rate per unit weight is highest for the immature fruit or vegetable and then steadily declines with age (Figure 3.1).

A significant group of fruits that includes the tomato, mango, banana and apple shows a variation from the described respiratory pattern in that a pronounced increase in respiration coincides with ripening (Figure 3.1). Such an increase in respiration is known as a respiratory climacteric, and this group of fruits is known as the climacteric class of fruits. The intensity and duration of the respiratory climacteric, first described in 1925 for the apple, varies widely among fruit species, as depicted in Figure 3.4. The commencement of the respiratory climacteric coincides approximately with the attainment of maximum fruit size (Figure 3.1), and it is during the climacteric that all the other changes characteristic of ripening occur. The respiratory climacteric, as

**Figure 3.4**

Respiratory patterns of some harvested climacteric fruits. SOURCE J.B. Biale (1950) Postharvest physiology and biochemistry of fruits, *Ann. Rev. Plant Physiol.* 1: 183–206. (With permission.)

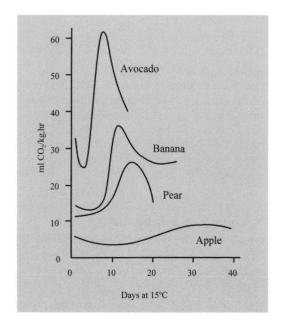

well as the complete ripening process, may proceed while the fruit is either attached to or detached from the plant (except for avocado, which will only ripen when detached from the plant).

Those fruits that do not exhibit a respiratory climacteric, such as citrus, pineapple and strawberry, are known as the non-climacteric class of fruit. Non-climacteric fruit exhibit most of the ripening changes, although these usually occur more slowly than those of the climacteric fruits. Table 3.2 lists some common climacteric and non-climacteric fruits. All vegetables can also be considered to have a non-climacteric type of respiratory pattern. The division of fruit into two classes on the basis of their respiratory pattern is an arbitrary classification but has served to stimulate considerable research to discover the biochemical control of the respiratory climacteric.

---

**TABLE 3.2 CLASSIFICATION OF SOME FRUIT ACCORDING TO THEIR RESPIRATORY BEHAVIOUR DURING RIPENING**

| CLIMACTERIC FRUIT | NON-CLIMACTERIC FRUIT |
|---|---|
| Apple (*Malus domestica*) | Cherry: sweet (*Prunus avium*) |
| Apricot (*Prunus armeniaca*) | sour (*Prunus cerasus*) |
| Avocado (*Persea americana*) | Cucumber (*Cucumis sativus*) |
| Banana (*Musa sp.*) | Grape (*Vitis vinifera*) |
| Blueberry (*Vaccinium corymbosum*) | Lemon (*Citrus limon*) |
| Cherimoya (*Annona cherimola*) | Pineapple (*Ananas comosus*) |
| Feijoa (*Feijoa sellowiana*) | Satsuma mandarin (*Citrus unshu*) |
| Fig (*Ficus carica*) | Strawberry (*Fragaria sp.*) |
| Kiwi fruit (*Actinidia deliciosa*) | Sweet orange (*Citrus sinensis*) |
| Mango (*Mangifera indica*) | Tamarillo (tree tomato) |
| Muskmelon (*Cucumis melle*) | (*Cyphomandra betacea*) |
| Papaya (*Carica papaya*) | |
| Passionfruit (*Passiflora edulis*) | |
| Peach (*Prunus persica*) | |
| Pear (*Pyrus communis*) | |
| Persimmon (*Diospyros kaki*) | |
| Plum (*Prunus sp.*) | |
| Tomato (*Lycopersicon esculentum*) | |
| Watermelon (*Citrullus lanatus*) | |

## Effect of ethylene

Climacteric and non-climacteric fruits may be further differentiated by their response to applied ethylene and by their pattern of ethylene production during ripening. It has been clearly established that all fruit produces minute quantities of ethylene during development. However, climacteric fruits produce much larger amounts of ethylene during ripening than non-climacteric fruits. This difference between the two classes of fruit is further exemplified by the internal ethylene concentration found at several stages of development and ripening (Table 3.3). The internal ethylene concentration of climacteric fruits varies widely, but that of non-climacteric fruits changes little during development and ripening. Ethylene, applied at a concentration as low as 0.1–1.0 µL/L for 1 day, is normally sufficient to hasten full ripening of climacteric fruit (Figure 3.5), but the magnitude of the climacteric is relatively independent of the concentration of applied ethylene. In contrast, applied ethylene merely causes a transient increase in the

TABLE 3.3  INTERNAL ETHYLENE CONCENTRATIONS MEASURED IN SEVERAL CLIMACTERIC AND NON-CLIMACTERIC FRUITS

| FRUIT | ETHYLENE (µL/L) |
|---|---|
| *Climacteric* | |
| Apple | 25–2500 |
| Pear | 80 |
| Peach | 0.9–20.7 |
| Nectarine | 3.6–602 |
| Avocado · | 28.9–74.2 |
| Banana | 0.05–2.1 |
| Mango | 0.04–3.0 |
| Passionfruit | 466–530 |
| Plum | 0.14–0.23 |
| Tomato | 3.6 –29.8 |
| *Non-climacteric* | |
| Lemon | 0.11–0.17 |
| Lime | 0.30–1.96 |
| Orange | 0.13–0.32 |
| Pineapple | 0.16–0.40 |

SOURCE From S.P Burg and E.A. Burg (1962) The role of ethylene in fruit ripening, *Plant Physiol.* 37: 179–89. (With permission.)

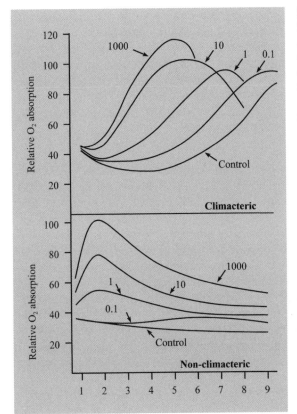

**Figure 3.5**
Effects of applied ethylene on respiration of climacteric and non-climacteric fruits. SOURCE J.B. Biale (1964) Growth, maturation, and senescence in fruits, *Science* 146: 880–88. (With permission.)

respiration of non-climacteric fruits, the magnitude of the increase depending on the concentration of ethylene (Figure 3.5). Moreover, the rise in respiration in response to ethylene may occur more than once in non-climacteric fruits in contrast to the single respiration increase in climacteric fruits.

The significance of ethylene for fruit ripening was established during the early part of the 20th century, when heaters burning kerosene were used to degreen — or colour yellow — California lemons. Denny, in 1924, found that while warmth was needed, the real cause of degreening was ethylene, and many workers soon demonstrated that ethylene could hasten the ripening of many fruits. Ethylene was regarded as an external agent that could promote the ripening of fruits, but in 1934 Gane found that fruit and other plant tissues produced extremely small quantities of ethylene. This finding was achieved despite the use of insensitive and time-consuming methods of analysis, which generally relied on trapping the ethylene with mercuric perchlorate, releasing the trapped ethylene and measuring it by manometry.

Research into the involvement of ethylene in fruit ripening was greatly stimulated by the development of sensitive gas chromatographic techniques for the measurement of low levels of ethylene, so that it is now possible to measure quantitatively as little as 0.001 μL/L in a 1 mL gas sample (see Appendix IV). This obviates the need for collecting or absorbing ethylene over a lengthy period of time.

## Ethylene biosynthesis

Ethylene has been shown to be produced from methionine via a pathway that includes the intermediates S-adenosyl-methionine (SAM) and 1-aminocyclopropane-1-carboxylic acid (ACC). The conversion of SAM to ACC by the enzyme ACC synthase is thought to be the rate limiting step in the biosynthesis of ethylene. However, in higher plants, ACC can be removed by conjugation to form malonyl ACC or glutamyl ACC. The addition of ACC to preclimacteric (unripe) fruit generally results in only a small increase in ethylene evolution, showing that another enzyme, the ethylene-forming enzyme (EFE or ACC oxidase), is required to convert ACC to ethylene. ACC oxidase is a labile enzyme that is sensitive to oxygen. Factors that affect the activity of ACC synthase include fruit ripening, senescence, auxin, physical injuries and chilling injury. This enzyme is believed to be a pyridoxal enzyme because it requires pyridoxal phosphate for maximal activity and is strongly inhibited by aminooxyacetic acid (AOA), rhizobitoxine and its analogue and L-2-amino-4-(2-aminoethoxy)-trans-3-butenoic acid (AVG), which are known inhibitors of pyridoxal phosphate-dependent enzymes. ACC oxidase is inhibited by anaerobiosis, temperatures above 35°C and cobalt ions. Small amounts of ethylene can also be formed in plant tissues from the oxidation of lipids involving a free-radical mechanism.

## Mode of action

Ethylene is a plant hormone that acts in concert with other plant hormones (auxins, gibberellins, kinins and abscisic acid) to exercise control over the fruit ripening process. Most is known about the relation of ethylene to fruit ripening because the availability of the sensitive gas chromatographic method for measurement of ethylene has enabled detailed studies of this relationship. The relationship of the other plant hormones to ripening is as yet not clearly defined.

It has been proposed that two systems exist for the regulation of ethylene biosynthesis. System 1 is initiated or perhaps controlled by an unknown factor that is probably involved in the regulation of senescence. System 1 then triggers system 2, which is responsible, during ripening of climacteric fruits, for the production of the large amounts of ethylene that are necessary for the full integration of ripening. System 2 is an autocatalytic process, with the production of ethylene triggering further production. Non-climacteric fruits do not have an active system 2, and treatment of climacteric fruits with ethylene circumvents system 1.

As in the case of other plant hormones, ethylene is believed to bind to specific receptor(s) to form a complex that then triggers ripening. Ethylene's action can be affected by altering the amount of receptor(s) or by interfering with the binding of ethylene to its receptor. Detailed studies of the structural requirements for biological activity of ethylene receptors led to the proposal that binding takes place reversibly at a site containing a metal, possibly copper. From kinetic studies on the responses of plant tissue to added ethylene it has been proposed that the affinity of the receptor for ethylene is increased by the presence of oxygen and decreased by carbon dioxide. The occurrence of a metal-containing receptor has not been confirmed but the proposition is supported by studies with silver ion. Treatment of fruit, flowers and other tissues with silver ion has been shown to inhibit the action of ethylene. The need for specific structural requirements for ethylene action has been demonstrated by treating tissues with analogues and antagonists of ethylene. The gaseous cyclic olefines, 2,5-norbornadiene and 1-methylcyclopropene (1-MCP), have been shown to be highly effective inhibitors of ethylene action. 1-MCP has been shown to bind irreversibly to the ethylene receptors in sensitive plant tissues and a single treatment with low concentrations for a few hours at ambient temperatures confers protection against ethylene for several days.

The pattern of changes in ethylene production rates and the internal concentrations of ethylene in relation to the onset of ripening have been observed in several climacteric fruits. In one type of fruit (e.g. banana, tomato and honey dew melon), the ethylene concentration rises before the onset of ripening, which was defined as the initial respiratory increase. In the second type (e.g. apple, avocado and mango), ethylene does not rise before the increase in respiration. In honey dew melon, the internal ethylene concentration rises from the preclimacteric level of 0.04 µL/L to 3.0 µL/L, at which concentration the fruit commences to ripen. The low concentrations of ethylene present in unripe fruit and the evident involvement of system 2 ethylene in ripening indicate that treatments that prevent ethylene from reaching a triggering concentration should delay ripening. Ripening has been delayed in green banana fruit for up to 180 days at 20°C when the fruit were ventilated continuously with an atmosphere of 5 per cent carbon dioxide, 3 per cent oxygen, and 92 per cent nitrogen.

It is well known that as many fruits develop and mature, they become more sensitive to ethylene. For some time after anthesis (flowering), young fruit can have high rates of ethylene production. Early in the life of fruit the concentration of applied ethylene required to initiate ripening is high, and the length of time to ripen is prolonged but decreases as the fruit matures (Table 3.4). The tomato is an extreme example of tolerance to ethylene. Banana and melons, in contrast, can be readily ripened with ethylene even when immature. Little is known about the factor(s) that control the sensitivity of the tissue to ethylene.

The earlier concept that an initial triggering of ripening by a single dose of ethylene is sufficient to ensure ripening is not now the case since application of silver ions, which block the ethylene receptor, will not only block the initiation of ripening by exogenous ethylene but also will arrest the ripening process during its progress. For example, colour development and enzyme synthesis will cease. Furthermore, one of the effects of storage under modified atmospheres (see Chapter 6) is for levels of ethylene produced by the fruit to diminish along with changes in colour and texture changes, while changes in sugars and acids

## TABLE 3.4 EFFECT OF MATURITY ON THE TIME TO RIPEN FOR TOMATOES

| | DAYS TO RIPEN | |
| MATURITY AT HARVEST (DAYS AFTER ANTHESIS) | TREATED WITH ETHYLENE | CONTROL |
| --- | --- | --- |
| 17 | 11 | ★ |
| 25 | 6 | – |
| 31 | 5 | 15 |
| 35 | 4 | 9 |
| 42 | 1 | 3 |

★ Had failed to ripen when experiment was terminated.

Note: Time to ripen was determined between anthesis and the first detectable red colour (first colour stage). Fruit were treated continuously with 1000 µL/L ethylene.

SOURCE J.M. Lyons and H.K. Pratt (1964) Effect of stage of maturity and ethylene treatment on respiration and ripening of tomato fruits, *Proc. Am. Soc. Hort. Sci.*, 84: 491–500. (With permission.)

responsible for some of the flavour proceed normally. It is apparent, therefore, that ethylene is only one of the regulatory components involved in ripening.

Ripening has long been considered to be a process of senescence and to be due to a breaking down of the cellular integrity of the tissue. Some ultrastructural and biochemical evidence supports this view. It is widely accepted that ripening is a programmed phase in the development of plant tissue, with altered nucleic acid and protein synthesis occurring at the commencement of the respiratory climacteric resulting in new or enhanced biochemical reactions operating in a coordinated manner. Both views fit with the known degradative and synthetic capacities of fruit during ripening. In view of the ample evidence of the ability of ethylene to initiate biochemical and physiological events, it is evident that ethylene action is regulated at the level of gene expression.

## Genetic control of ripening

It will be no surprise that ripening is under genetic control in the cells of the fruit. In recent years, research has begun to unravel the complex process of the genetic control of ripening. There is good evidence that there is a marked increase in protein synthesis and nucleic acid synthesis (especially messenger RNA synthesis) preceding and during the early stages of the respiratory climacteric for climacteric fruits such as avocado, banana and tomato. Gel electrophoresis has been used to compare the presence of certain proteins during fruit growth and ripening, and has demonstrated that the levels of some proteins increase during ripening and that new types of proteins are synthesised during the ripening process. Similarly, mRNAs increase during ripening, and *in vitro* translation of these mRNAs can be shown to produce the new proteins. All this evidence supports the view that ripening is under genetic control.

Additional evidence comes from a series of natural mutants of tomato and variants produced by genetic transformation that exhibit abnormal ripening behaviour. A number of non-ripening or slow ripening mutants have been identified that appear to affect the processes of ethylene synthesis, ethylene perception and/or signal transduction leading to abnormal colouring (often yellow or pale red), lack of softening or low ethylene production (see Table 3.5). These mutants have been extensively used to determine some of the fundamental processes involved in ripening, particularly softening. A number of different hydrolase enzymes break down the carbohydrate polymers (e.g. pectins, celluloses, hemicelluloses) responsible for the structural integrity of cell walls. Among these is the enzyme polygalacturonase (PG), which hydrolyses the $\alpha(1\text{-}4)$ linkage between galacturonic acid residues in pectins. Early research with the tomato mutants suggested this could be the primary enzyme responsible for softening. However, as with many systems, further work has shown that softening is more complex, probably also involving a number of other mechanisms including the hydrolase enzyme, pectinesterase, which demethylates galacturonic acid residues in pectins, the release of $Ca^{2+}$, which is important in cross-linking the polymer chains in the cell wall, and the consequent swelling in the middle lamella of the adjacent cell walls, allowing the cells to move apart. This gives the softer mouthfeel characteristic of a ripe fruit.

## TABLE 3.5  MUTANTS OF TOMATO WITH ABNORMAL RIPENING BEHAVIOUR

| MUTANT | LOCATION ON CHROMOSOME | FRUIT PHENOTYPE COMPARED WITH WILD TYPE |
|---|---|---|
| Ripening inhibitor (*rin*) | 5 | normal growth, slowly turns pale yellow; very low ethylene production; little softening, very low PG activity; does not ripen after exposure to ethylene; high oxygen causes light pink colour |
| Never ripe (*Nr*) | 9 | normal growth, slowly turns orange-red; limited softening, ethylene production, PG and lycopene synthesis |
| Non ripening (*nor*) | 10 | more extreme than *rin*; final colour is deep yellow; very low ethylene production; contains <1% of wild type PG; high NaCl causes faster ripening, deep orange colour and some softening |

SOURCE Adapted from G. Hobson and D. Grierson, G.B. Seymour, J.E. Taylor and G.A. Tucker (eds) (1993) *Biochemistry of fruit ripening*, Chapman & Hall, London. pp. 405–42.

# BIOCHEMISTRY OF RESPIRATION

All living organisms require a continuous supply of energy. This energy enables the organism to carry out the necessary metabolic reactions to maintain cellular organisation, to transport metabolites around the tissue and to maintain membrane permeability. In addition, a continuous supply of the organic molecules needed for synthetic reactions in the cells is required.

## *Aerobic metabolism*

Most of the energy required by horticultural produce is supplied by aerobic respiration, which involves the oxidative breakdown of certain organic substances stored in the tissues. A common substrate for respiration is glucose and, if it is completely oxidised, the overall reaction is:

$$C_6H_{12}O_6 + 6O_2 \rightarrow 6CO_2 + 6H_2O + energy$$

Respiration is essentially the reverse of photosynthesis, by which energy derived from the sun is stored as chemical energy, mainly in carbohydrates containing glucose. Full utilisation of glucose involves two main reaction sequences in respiration:

1. glucose $\rightarrow$ pyruvate by the enzymes of the Embden-Meyerhof-Parnas (EMP) pathway, or glycolysis, which are located in the cytoplasm;

2. pyruvate $\rightarrow$ carbon dioxide by the tricarboxylic acid (TCA) cycle, the enzymes of which are located in the mitochondria.

Free glucose is conventionally the compound involved in the initial oxidative step, but it is not the storage form of carbohydrate in the plant. Starch, a polymer of glucose, is often the main storage carbohydrate, and it must be degraded first to glucose by enzymes such as the amylases and maltase or to glucose-1-phosphate by the enzyme phosphorylase. Some commodities have a high sucrose content, which can be hydrolysed to glucose and fructose by the enzyme invertase or via sucrose synthase to UDP-glucose and so to glucose-1-phosphate. Interconversion of sucrose and starch is also possible in many plant tissues. Figure 3.6 provides a generalised scheme for the initial conversions of the storage carbohydrates. Glucose and fructose released from starch and sucrose are oxidised to other respiratory substrates (Figure 3.7).

## *Glycolysis (EMP sequence)*

A simplified view of glycolsis is shown in Figure 3.7. The overall reaction system can be balanced as:

glucose + 2ADP + 2Pi + 2NAD $\rightarrow$ 2pyruvate + 2ATP + 2NADH + 2H$_2$O

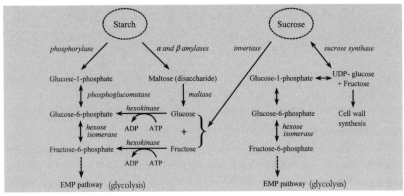

**Figure 3.6**
Starch and sucrose degradation to form the 6-carbon sugars, glucose and fructose, for oxidation via glycolysis. Some of the enzymes involved are indicated, as are the phosphorylation reactions of glucose and fructose. Sucrose may also be degraded to UDP-glucose (uridine diphosphoglucose), a key nucleotide sugar involved in cell wall synthesis, starch/sucrose and other important carbohydrate interconversions.

**Figure 3.7**
A simplified scheme for aerobic respiration of carbohydrate reserves in plants via glycolysis and the TCA cycle. Malate from the vacuole or mitochondria can be directly metabolised to pyruvate by NAD- or NADP-specific malic enzyme (ME). Pyruvate produced by ME or from fructose 6-phosphate can be oxidised via the TCA cycle or by anaerobic respiration. The conversion of intermediates in these pathways to $CO_2$ and the reduction of NAD and FAD to NADH and $FADH_2$, respectively, are also indicated. NADH and $FADH_2$ are oxidised via the electron transport chain (see Figure 3.8).

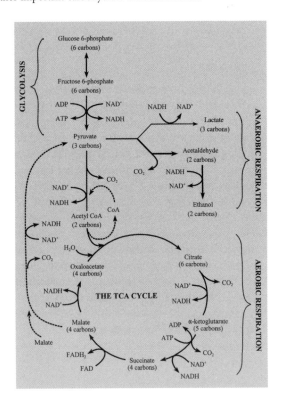

The energy liberated from the reaction system is trapped and stored in adenosine triphosphate (ATP) and reduced nicotinamide adenine dinucleotide (NADH), with oxidation of each molecule of NADH giving 3ATP (Figure 3.8). The total energy liberated by the conversion of glucose to pyruvate is, therefore, equivalent to 8ATP. The energy is subsequently made available to the plant by breaking a phosphate bond in the reverse reaction:

$$ATP \rightarrow ADP + Pi + energy$$

This energy can then be used in a wide range of synthetic reactions and metabolic interconversions in the plant.

## TCA cycle (tricarboxylic acid cycle)

A simplified view of the full cycle is shown in Figure 3.7. The overall reaction system is:

$$pyruvate + 2.5O_2 + 15ADP + 15Pi \rightarrow 3CO_2 + 2H_2O + 15ATP$$

The energy of the original glucose molecule (giving 2x pyruvate) liberated from the TCA cycle is 30ATP (compared with 8ATP from the EMP sequence). The carbon dioxide produced in respiration is derived from the TCA cycle under aerobic conditions and involves a consumption of oxygen. The rate of respiration can therefore be measured by the amount of carbon dioxide produced or oxygen consumed.

The total chemical energy liberated during the oxidation of 1 mole of glucose is approximately 1.6 MJ. About 90 per cent of this energy is preserved within the plant system, and the remainder is lost as heat. Respiration is therefore an efficient converter of energy compared with man-made devices for energy conversion. For example, in the petrol engine more than 50 per cent of the liberated energy is lost as heat.

**Figure 3.8**
NADH and FADH$_2$ are oxidised via the electron transport chain to produce water and ATP from ADP. Three ATP molecules are produced from NADH and two from FADH$_2$.

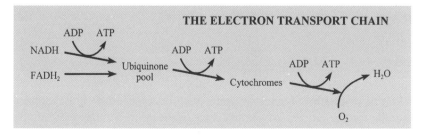

## Other respiratory pathways

The oxidative pentose phosphate pathway (OPPP) converts glucose-6-phosphate to fructose-6-phosphate and glyceraldehyde-3-phosphate and $CO_2$ through a complex cyclic reaction pathway involving 4-, 5-, 6- and 7-carbon sugar phosphates (Figure 3.9). Although it is not considered to be a major respiratory pathway in fruits, the OPPP does provide ribose-5-phosphate for nucleotide and nucleic acid synthesis, erythrose-4-phosphate for shikimic acid and aromatic amino acid biosynthesis and NADPH to drive a variety of synthetic reactions. The OPPP is of greater importance in leafy vegetables and ornamentals in which it can account for a significant proportion of tissue respiration, perhaps in the range of 10–20 per cent.

The vacuoles of many fruit and vegetables contain high concentrations of organic acids, particularly malic and citric acids, which can be used as respiratory substrates. These acids can be utilised directly by the TCA cycle in the mitochondria. Malic acid can also undergo reductive decarboxylation with the evolution of carbon dioxide and production of NADPH or NADH and pyruvate by malic enzyme present in the cytosol or in the mitochondria (Figure 3.7).

## Electron transport system and oxidative phosphorylation

The NADH and $FADH_2$ produced as a result of certain reactions in glycolysis and the TCA cycle are oxidised through the electron transport system, which is a complex set of dehydrogenase enzymes, electron carrier compounds such as ubiquinones and cytochromes, and cytochrome oxidase bound in the mitochondrial membranes (Figure 3.8). This process utilises oxygen in the final step catalysed by cytochrome oxidase, and is responsible for the physiologically measurable uptake of oxygen. The system is tightly linked to the production of ATP (from ADP and Pi), and hence is termed oxidative phosphorylation.

A number of other oxidase enzymes are present in plants and their fruit tissues. These include polyphenol oxidases, ascorbate oxidases and others, but their role in respiration is unclear, though they do contribute to the oxygen uptake of plant tissues to varying degrees.

## Respiratory quotient (RQ)

The complete oxidation of malate

$$C_4H_6O_5 + 3O_2 \rightarrow 4CO_2 + 3H_2O$$

generates more carbon dioxide than the amount of oxygen consumed, whereas oxidation of glucose generates an equal amount of carbon dioxide for the oxygen consumed. This relationship becomes important when measuring respiration by gas exchange, in which the carbon dioxide evolved and/or oxygen consumed is measured (i.e. it is possible to record different values for respiration depending on which gas is monitored). Ideally both gases should be measured simultaneously.

The concept of respiratory quotient (RQ) has been developed to quantify this variation, where:

$$RQ = CO_2 \text{ produced (mL)}/O_2 \text{ consumed (mL)}$$

For the complete oxidation of glucose, RQ = 1.0, whereas for malate RQ = 1.3.

An alternative substrate could be long-chain fatty acids, for example, stearic acid:

$$C_{18}H_{36}O_2 + 26O_2 \rightarrow 18CO_2 + 18H_2O$$

These fatty acids have much less oxygen per carbon atom than sugars and, therefore, require a greater oxygen consumption for the production of carbon dioxide. The RQ equals 0.7 for the above reaction.

Measurement of RQ itself can give some guide to the type of substrate that is being respired: a low RQ suggests some fat metabolism and a high RQ suggests organic acids. Changes in RQ during growth and storage can also indicate a change in the type of substrate that is being metabolised.

**Figure 3.9**

Oxidative Pentose Phosphate Pathway. This diagram illustrates one of several variants of this pathway, located in the cytosol.

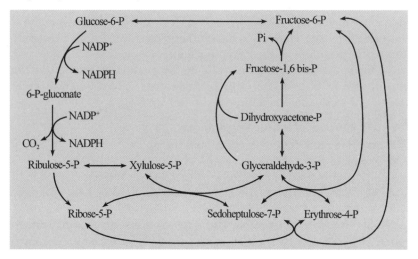

## Anaerobic metabolism (fermentation)

The aerobic respiratory pathways utilise oxygen and are the preferred pathways. The normal atmosphere is rich in oxygen, so the amount of oxygen available in the tissue is unlimited. Under various storage conditions the amount of oxygen in the atmosphere may be limited and insufficient to maintain full aerobic metabolism. Under these conditions the tissue can initiate anaerobic respiration, by which glucose is converted to pyruvate by the EMP pathway. But pyruvate is then metabolised into either lactic acid or acetaldehyde and ethanol in a process termed fermentation (Figure 3.7). The oxygen concentration at which anaerobic respiration commences varies between tissues and is known as the extinction or anaerobic compensation point. The oxygen concentration at this point depends on several factors, such as species, cultivar, maturity and temperature. Anaerobic respiration produces much less energy per mole of glucose than aerobic pathways, but it does allow some energy to be made available to the tissue under adverse conditions. An increasing RQ is generally indicative of a switch to fermentation reactions. Anaerobic fermentation will usually lead to production of off-flavours and odours in fruit and vegetables.

## Metabolites for synthetic reactions

The respiratory pathways are not only used for the production of energy for the tissue. Carbon skeletons are required for many synthetic reactions in the cell, and these skeletons can be removed at several points. For example, α-ketoglutarate may be converted to the amino acid glutamate, from which several other amino acids may be produced for protein synthesis; succinate may be diverted into the synthesis of various heme pigments including chlorophyll. The loss of α-ketoglutarate and succinate from the TCA cycle for synthetic reactions would eventually lead to the stopping of the cycle. Therefore, C4 acids are fed into the cycle. These are produced principally by the fixation of carbon dioxide into phosphoenol-pyruvate to give oxaloacetate. Alternatively, vacuolar reserves of malate, for example, may be utilised.

## Genetic control of plant metabolism

The environmental control of plant metabolism has been used for a long time in a more or less empirical fashion by the use of low temperature. For example, to store fruits and vegetables and to bring about rapid cooling of leafy vegetables, in particular, after harvest from the field. Cooling markedly reduces the rate of respiration of the tissues, generally prolonging their useful life (Chapter 4). Controlled atmosphere (CA) and modified atmosphere (MA) storage of fruits and vegetables (Chapter 6), using elevated concentrations of carbon dioxide and lowered concentrations of oxygen compared to air, also are used to slow respiratory processes in produce.

More subtle means of manipulating the postharvest behaviour of fruits and vegetables are now becoming available by using mutant varieties or genetic manipulation to develop transgenic plants in which particular enzymes (gene products) can be either virtually eliminated or enhanced. A notable example is the reduction in the activity of the enzyme polygalacturonase, which is involved in the breakdown of the cell wall during ripening of the tomato. This research has resulted in the commercial production and marketing of the FLAVR SAVR™ tomato, which remains firmer longer during ripening so that it can be kept on the plant longer before picking. The object of this is to enhance the flavour compared with tomatoes picked at an early stage of colouring to enable the fruit to withstand the mechanical handling during transport and marketing. Transgenic fruits containing ACC deaminase and antisense ACC synthase, ACC oxidase and polyphenoloxidase have also been produced. The first three transformations reduce ethylene production and slow ripening, while lowered polyphenoloxidase activity reduces browning of damaged tissue.

Apart from any concerns that consumers may have about eating genetically manipulated produce, these genetic techniques will need to be approached with some caution since plant metabolism is very adaptable due to the fact that there are usually alternate pathways by which a given metabolic product or intermediary metabolite may be produced. This plasticity of metabolism makes plants and their products able to modify their performance under a wide variety of conditions but will make the task of the genetic engineer more difficult. This is borne out by very recent studies in which particular enzyme levels have been modified in plants by genetic manipulation in order to study metabolic control. Formerly it was thought that a small number of key enzymes in each pathway regulated the flow of carbon through a particular pathway. This was thought to be achieved by low absolute concentrations of the regulatory enzyme, with its activity depending on one or more co-factors such as the ratio of oxidised to reduced nicotinamide nucleotides (NAD/NADH; NADP/NADPH) or phosphorylating co-factor (ADP/ATP). However, the virtual removal of a particular enzymic activity by genetic manipulation is found to have little effect in a number of cases, so a combination of metabolic plasticity (mentioned above) and much more dispersed regulatory control mechanisms must be invoked.

# CHEMICAL CHANGES DURING MATURATION
.......

At some stage during the growth and development of fruit and vegetables, the consumer recognises that the produce has attained optimum eating condition. This desirable quality is not associated with any universal change, but is attained in various ways in different tissues (see Chapter 10).

## Fruit

Climacteric fruit generally reach the fully ripe stage after the respiratory climacteric. However, it is the other events initiated by ethylene that the consumer associates with ripening.

### Colour

Colour is the most obvious change that occurs in many fruits and is often the major criterion used by consumers to determine whether the fruit is ripe or unripe. The most common change is the loss of green colour. With a few exceptions — such as the avocado, kiwifruit and Granny Smith apple — climacteric fruits show rapid loss of green colour on ripening. Many non-climacteric fruits also exhibit a marked loss of green colour when they reach optimum eating quality, such as citrus fruit in temperate climates (but not in tropical climates). The green colour is due to the presence of chlorophyll, which is a magnesium–organic complex. The loss of green colour is due to degradation of the chlorophyll structure. The principal agents responsible for this degradation are pH changes (mainly due to leakage of organic acids from the vacuole), oxidative systems and chlorophyllases (Figure 3.10). Loss of colour depends on one or all of these factors acting in sequence to destroy the chlorophyll structure.

The disappearance of chlorophyll is often associated with the synthesis and/or revelation of pigments ranging from yellow to red. Many of these pigments are carotenoids, which are unsaturated hydrocarbons with generally 40 carbon atoms and which may have one or more oxygen atoms in the molecule. Carotenoids are stable compounds and remain intact in the tissue even when extensive senescence has occurred. Carotenoids may be synthesised during the development stages on the plant but remain masked by the presence of chlorophyll. Following the degradation of chlorophyll, the carotenoid pigments become visible. With other tissues, carotenoid synthesis occurs concurrently with chlorophyll degradation. Banana peel is an example of the former system and tomato of the latter.

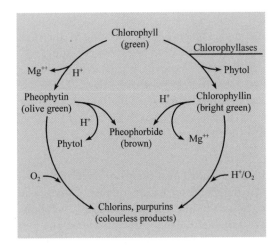

**Figure 3.10**
Some pathways for the degradation of chlorophyll.

Anthocyanins provide many of the red-purple colours of fruit, vegetables and flowers. Anthocyanins are water-soluble phenolic glucosides that can be found in the cell vacuoles of fruit and vegetables (e.g. beetroot), but are often in the epidermal layers (e.g. apples and grapes). They produce strong colours, which often mask carotenoids and chlorophyll.

*Carbohydrates*

The largest quantitative change associated with ripening is usually the breakdown of carbohydrate polymers, especially the near total conversion of starch to sugars. This alters both the taste and texture of the produce. The increase in sugar renders the fruit much sweeter and, therefore, more acceptable. Even with non-climacteric fruits, the accumulation of sugar is associated with the development of optimum eating quality, although the sugar may be derived from sap imported into the fruit rather than from the breakdown of the fruit's starch reserves.

The breakdown of polymeric carbohydrates, especially pectic substances and hemicelluloses, weakens cell walls and the cohesive forces binding cells together. In the initial stages, the texture becomes more palatable, but eventually the plant structures disintegrate. Protopectin is the insoluble parent form of pectic substances. In addition to being a large polymer, it is cross-linked to other polymer chains with calcium bridges and is bound to other sugars and phosphate derivatives to form an extremely large polymer. During ripening and maturation, protopectin is gradually broken down to lower molecular weight fractions, which are more soluble in water. The rate of degradation of pectic substances is directly correlated with the rate of softening of fruit.

*Organic acids*

Usually organic acids decline during ripening as they are respired or converted to sugars. Acids can be considered as a reserve source of energy to the fruit, and would therefore be expected to decline during the greater metabolic activity that occurs with ripening. There are exceptions, such as bananas, where the highest level is attained when the banana is fully ripe, but the acidic level is not high at any stage of development compared with other produce.

*Nitrogenous compounds*

Proteins and free amino acids are minor constituents of fruit and, as far as is known, have no role in determining eating quality. Changes in nitrogenous constituents do, however, indicate variations in metabolic activity during different growth phases. During the climacteric phase of many fruits there is a decrease in free amino acids, which often reflects an increase in protein synthesis. During senescence, the level of free amino acids increases, reflecting a breakdown of enzymes and decreased metabolic activity.

*Aroma*

Aroma plays an important part in the development of optimal eating quality in most fruit. Aroma is due to the synthesis of many volatile organic compounds (often known merely as volatiles) during the ripening phase. The total amount of carbon involved in the synthesis of volatiles is less than 1 per cent of that expelled as carbon dioxide. The major volatile formed is ethylene, which accounts for about 50–75 per cent of the total carbon in the volatiles; ethylene does not contribute to typical fruit aromas. The amount of aroma compounds is therefore extremely small. Chapter 2 discussed the nature of the compounds formed. Non-climacteric fruits also produce volatiles as they reach optimum eating quality. These fruits do not synthesise compounds that are as aromatic as those in climacteric fruit; nevertheless, the volatiles produced are still appreciated by consumers.

## Vegetables

Vegetables generally show no sudden increase in metabolic activity that parallels the onset of the climacteric in fruit, unless sprouting or regrowth is initiated. The process of germination is sometimes deliberately applied to some seeds, such as mung bean, and the sprouted product is the marketed vegetable. Apart from obvious anatomical changes during sprouting, considerable compositional changes occur. The sugar level increases markedly as the result of the rapid conversion of fats or starch. From a nutritional view, the increase in vitamin C in sprouted seeds can be valuable in diets with marginal vitamin C intakes.

Vegetables can be divided into three main groups: seeds and pods; flowers, buds, stems and leaves; and bulbs, roots and tubers. Some fruits are also consumed as vegetables; they may be either ripe (e.g. tomato, egg-plant) or immature (e.g. zucchini, cucumber, okra).

Seeds and pods, if harvested when fully mature, as is the practice with cereals, have low metabolic rates because of their low water content. In contrast, all seeds consumed as fresh vegetables, such as legumes and sweet corn, have high levels of metabolic activity because they are harvested at an immature stage, often with the inclusion of non-seed material, such as bean pod (pericarp). Eating quality is determined by flavour and texture and not by physiological age. Generally the seeds are sweeter and more tender at an immature stage. With advancing maturity, the sugars are converted to starch with the resultant loss of sweetness, the water content decreases and the amount of fibrous material increases. Seeds for consumption as fresh produce are harvested when the water content is about 70 per cent; in contrast, dormant seeds are harvested at less than 15 per cent water.

Edible flowers, buds, stems and leaves vary greatly in metabolic activity and hence in the rate of deterioration. Stems and leaves often senesce rapidly and therefore lose their attractiveness and nutritional value. Generally the most visible sign of senescence is degreening, resulting in yellowing due to underlying carotenoid pigments. Texture often becomes the dominant characteristic that determines both the harvest date and quality, with reduced turgor through water loss altering the texture. The natural flavour is often of less importance than texture, as many of these vegetables are cooked with salt or spices added. Growth processes such as cell division and expansion and protein and carbohydrate synthesis usually cease, and the metabolism goes into a catabolic or degradative mode.

Bulbs, roots and tubers are storage organs that contain food reserves required when growth of the plant is resumed (they are often held for the purpose of propagation). When harvested, their metabolic rate is low and, under appropriate storage conditions, their dormancy can be prolonged. The biochemistry of these storage organs is geared to a slow metabolic rate designed to provide the low levels of energy required to maintain life in the cells of these tissues during dormancy. Postharvest considerations are to maintain produce in the dormant state.

## Ornamentals

Ornamentals can be grouped into cut flowers, cut foliage and pot plants. Others, of course, are planted in gardens and urban landscaping, but they are beyond the scope of this book.

Flowers usually have high rates of respiration through glycolysis and the TCA cycle based on sugar translocation from the leaves. For cut flowers, therefore, so-called preservative solutions, used by florists, contain sucrose as a carbohydrate source to help maintain the respiration rate and to extend storage life.

### Pigments

The main pigments in ornamental foliage are similar to those in fruit and vegetables. Chlorophyll is the principle green foliage pigment and carotenoids constitute many of the yellow, orange and red pigments. However, the anthocyanins and related compounds are responsible for red, purple and blue colours in most flowers. The observed colours are largely related to the pH of the flower sap. Anthocyanins are red at a more acidic pH (below 7), whereas they tend to be blue above pH 7. This gives rise to the phenomenon in roses known as 'blueing', where a shift from red to blue colouration occurs with ageing. This is due to the depletion of sugars as a respiratory substrate and the switch to catobolism of proteins, with the release of free amino groups resulting in a shift to a more more alkaline pH in the cell sap. The 'colour' white is due to a total reflectance of the visible spectrum and often results from the presence of highly aerated tissues, as in some flowers. Variegated leaves in some ornamentals, for example, are due to areas devoid of chloroplasts and may be white or orange/yellow, the latter colour due to the carotenoid pigments.

### Ethylene

Flowers tend to have a short postharvest (vase) life. A major contributing factor is often attributed to their sensitivity to ethylene. Most ornamentals should be regarded as non-climacteric, although some produce a distinct ethylene peak and respiratory climacteric. There will therefore be differential responses to ethylene similar to fruit and vegetables. Thus some flowers, such as the non-climacteric delphinium, are highly sensitive to ethylene while the climacteric carnation is relatively tolerant. The major effect of ethylene is induced abscission. The short postharvest life of cut flowers is, however, confounded by limited carbohydrate reserves and an overall rapid rate of metabolism and development.

# FURTHER READING
······

Abeles, F.B., P.W. Morgan and M.E. Saltveit, Jr (1992) *Ethylene in plant biology* (2nd ed.), Academic Press, New York.

Borochov, A. and R. Woodson (1990) Physiology and biochemistry of flower petal senescence, *Hortic. Rev.* 11: 15–43.

Brady, C.J. (1987) Fruit ripening, *Annu. Rev. Plant Physiol.* 38: 155–78.

Burton, W.G. (1982) *Postharvest physiology of food crops*, Longman, New York.

Davies, D.D. (ed.) (1988) Biochemistry of metabolism, vol. 11, P.K. Stumpf and E.E. Conn (eds) *The biochemistry of plants*, Academic Press, London.

Davies, D.D. (ed.) (1988) Physiology of metabolism, vol. 12, P.K. Stumpf and E.E. Conn (eds) *The biochemistry of plants*, Academic Press, London.

Dennis, D.T. and D.H. Turpin (1990) *Plant physiology, biochemistry and molecular biology*, Longmand, London.

Douce, R., D.A. Day (eds) (1985) Higher plant cell respiration, vol. 18, *Encyclopaedia of Plant Physiology*, Springer-Verlag, Berlin.

Edwards, G. and D.A. Walker (1983) *C3, C4: Mechanisms and cellular and environmental regulation of photosynthesis*, Blackwell Scientific Publications, Oxford,

Fischer, R.L. and A.B. Bennett (1991) Role of cell wall hydrolases in fruit ripening, *Annu. Rev. Plant Physiol.* 42: 675–703.

Fluhr, R. and A.K. Mattoo (1996) Ethylene-biosynthesis and reception, *Critical Rev. Plant Sciences*, 15(5&6): 479–523.

Halevy, A.H. and S. Mayak (1979) Senescence and postharvest physiology of cut flowers, Part 1, *Hortic. Rev.* 1: 204–36.

Kende, H. (1993) Ethylene biosynthesis, *Annu. Rev. Plant Physiol.* 44: 283–307.

Klee, H. and M. Estelle (1991) Molecular genetic approaches to plant hormone biology, *Annu. Rev. Plant Physiol.* 42: 529–51.

Lea, P.J. and R.C. Leegood (eds) (1993) *Plant biochemistry and molecular biology*, Wiley, New York.

Morris, S.C. and B.J. Pogson (1997) *Postharvest senescence of vegetable crops and its regulation*, Gyanodaya Prakashan, Ed. Palni, L.M.S. Nainital, India.

Salunkhe, D.K. (1984) *Postharvest biotechnology of vegetables*, vols 1 and 2, CRC Press, Boca Raton, FL.

Salunkhe, D.K. and B.B. Desai (1984) *Postharvest biotechnology of fruits*, vols 1 and 2, CRC Press, Boca Raton, FL.

Seymour, G.B., J.E. Taylor and G.A. Tucker (eds) (1993) *Biochemistry of fruit ripening*, Chapman & Hall, London.

Stitt, M. and V. Sonnewald (1995) Regulation of metabolism in transgenic plants, *Annu. Rev. Plant Physiol.* 46: 341–68.

Taiz, L. and E. Zeiger (1991) *Plant physiology*, Benjamin/Cummins Publishing Co., Redwood City, CA.

Wardowski, W.F., S. Nagy and W. Grierson (1986) *Fresh citrus fruits*, AVI, Westport, CT.

Weichmann, J. (ed.) (1987) *Postharvest physiology of vegetables*, Marcel Dekker, New York.

# 4

# EFFECTS OF TEMPERATURE

Temperature is the single most important factor governing the maintenance of postharvest quality in fruits, vegetables and ornamentals. Temperature responses of harvested horticultural produce can be generally classed as:

◆ normal intermediate temperature range effects;
◆ adverse low temperature effects; and
◆ adverse high temperature effects (Figure 4.1).

In the case of adverse low temperature effects, clear mechanistic distinctions between freezing and chilling injuries can be made. Specific disorders and further details on the nature of low and high temperature effects are given in Chapters 8 and 13, respectively.

**Figure 4.1**
Responses of
non–chilling-sensitive
and chilling-sensitive
produce to temperature.

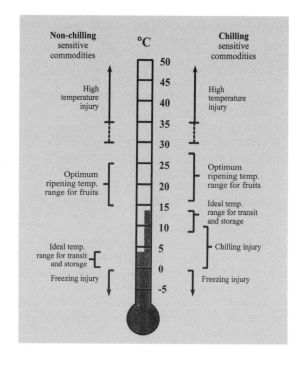

## Normal intermediate temperature effects

Harvested produce is ideally transported and stored under reduced temperatures likely to maximise longevity. However, the effect of reducing temperature on the maintenance of produce quality is not uniform over the normal intermediate or physiological temperature range (0–30°C for non-chilling-sensitive lines, 7.5–30°C for moderately chilling-sensitive lines and 13–30°C for chilling-sensitive lines). Only a small improvement in storage life — the time that produce can be held in an acceptable condition after harvest — is achieved by small reductions in temperature at the upper end of the temperature range. In contrast, much larger improvements are obtained by similarly small reductions at lower temperatures (Figure 4.1), where even a change in temperature of 1°C can have a significant effect. Ideally, the greatest reduction in processes associated with deterioration (e.g. respiration, change in texture, loss of vitamin C), and thus the best maintenance of quality, will be obtained if produce is held just above its freezing point temperature or just above its chilling threshold temperature in the case of chilling-sensitive produce.

Metabolism (e.g. respiration) in fruit, vegetables and ornamentals involves many enzyme reactions. The rate of these reactions within the physiological temperature range generally increases exponentially with an increase in temperature. This relationship may be described mathematically by use of the temperature quotient $Q_{10}$. Van't Hoff, a Dutch chemist, determined that the rate of a chemical reaction approximately doubles for each 10°C rise in temperature:

$$Q_{10} = (R_2/R_1)^{10/(t_2-t_1)} = \text{constant, about 2}$$

where $t_2$ and $t_1$ are the higher and lower temperatures (°C) and $R_2$ and $R_1$ are the respective metabolic rates. Using this relationship, either the $Q_{10}$ or an unknown rate for any temperature difference may be calculated. For many biological processes, $Q_{10}$ does not remain constant over the physiological range (Table 4.1). That is, $Q_{10}$ is a function of temperature. $Q_{10}$ values are generally, but not always, highest between 1°C and 10°C (e.g. seven), but at temperatures above 10°C fall to between two and three (Figure 4.2).

Lowering the temperature of both climacteric and non-climacteric produce lowers their rate of deterioration (i.e. high quality is retained longer and shelf life is increased) (Figure 4.2). However, in the case of climacteric fruit, low temperature can be used to achieve a delay in the onset of ripening. The effect of decreased temperature on ripening follows an exponential relationship similar to that shown in Figure 4.1.

## TABLE 4.1 RESPIRATION RATE AND HEAT PRODUCTION OF CARNATIONS AT DIFFERENT TEMPERATURES

| TEMPERATURE (°C) | RESPIRATION RATE (MG $CO_2$/KG/H) | HEAT PRODUCTION (KJ/TONNE/H)* | $Q_{10}$ |
|---|---|---|---|
| 0 | 10 | 104 | — |
| 10 | 30 | 320 | 3.0 |
| 20 | 239 | 2 550 | 8.0 |
| 30 | 516 | 5 504 | 2.2 |
| 40 | 1 053 | 11 232 | 2.0 |
| 50 | 1 600 | 17 126 | 1.5 |

* Conversion from BTU/ton/h.

SOURCE E.C. Maxie, D.S. Farnham, F.G. Mitchell, N.F. Sommer, R.A. Parsons, R.G. Snyder and H.L. Rae (1973) Temperature and ethylene effects on cut flowers of carnations (*Dianthus caryophyllus*), J. Am. Soc. Hort. Sci. 98(6): 568–72.

Lowering temperature not only reduces the production of ethylene, but also the rate of response of the tissues to ethylene. Thus, at lower temperature, longer exposure to a given concentration of ethylene is required to initiate ripening or enhance senescence.

Normal ripening occurs only within a particular range of temperature (commonly 10–30°C), although some fruit, such as some pear cultivars, will ripen slowly but satisfactorily at temperatures below 10°C. The best quality in fruit, however, generally develops at a ripening temperature of 20–23°C.

Provided that a fruit is not sensitive to chilling, maximum storage life can be achieved at temperatures below the ripening range. For example, Williams Bon Chretien (WBC) (or Bartlett) pears will not ripen at temperatures below about 12°C, and maximum storage life is obtained by storage at −1°C followed by removal to temperatures greater than 12°C when ripening is desired. However, if the pears are held too long at low, non-ripening temperatures, they will fail to ripen normally after they are exposed to ripening temperatures.

For chilling-sensitive produce, the best retention of quality is obtained just above the chilling threshold temperature. By comparison, non-chilling-sensitive produce should be stored and handled at just above the freezing temperature. In addition to delaying fruit ripening

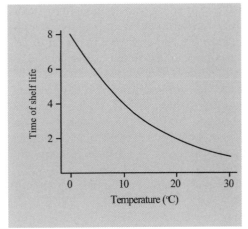

**Figure 4.2**
Illustration of a simple hypothetical $Q_{10}$ relationship in plants and the effect on quality retention (i.e. time of shelf life). The numbers on the vertical axis refer to $Q_{10}$ values that increase with lower temperatures, as does shelf life.

and senescence processes, these basic principles of temperature management can be applied to inhibit all manner of development processes in fresh produce, such as opening of cut flowers, toughening asparagus and loss of sweetness in peas.

## Adverse low temperature effects

Produce may be exposed to undesirably low temperatures for a number of reasons, such as transport in cold climatic regions, incorrect setting of the thermostat in storage rooms or exposure to sub-zero coolant (e.g. dry ice). Freezing injury occurs at temperatures of 0°C or below, and involves intercellular and/or intracellular ice formation. The precise temperature at which freezing occurs depends upon the concentration of solutes in the tissue, with the freezing point being lowered further with increasing osmotic concentration (i.e. freezing point depression). For example, lettuce freezes at just below 0°C (about −0.2°C), while grapes, which have a very high sugar content (ca. 14 per cent of fresh weight), do not freeze until less than −2.0°C.

Freezing of tissue water initiates desiccation and osmotic stress of cellular structures, such as membranes, and constituents, such as proteins, because solvent water is lost to support the growth of ice crystals. In addition, expansion of the water upon freezing, especially intracellular ice formation, can cause considerable physical disruption to the cell structure. Upon thawing, affected tissue usually cannot resume normal metabolism or regain normal texture. Adversely affected freeze-thawed tissue is flaccid and/or water-soaked. However, some produce — such as cabbage, onions and some pear cultivars — may be thawed without detriment. This regaining of normal form and function can be achieved if ice crystal damage is minimal and if the rate of temperature rise is sufficiently slow to allow orderly redistribution of water and reformation of intracellular compartmentalisation (i.e. the organised partition of cell functions within the membrane-bound compartments, or organelles, in the cell).

Chilling injury of susceptible commodities occurs at low temperatures that are above the freezing point of the produce. Injury is the result of imbalanced metabolism and loss of cellular compartmentalisation at suboptimal temperatures. Factors that influence susceptibility to chilling injury are discussed in Chapter 8, as are the myriad chilling injury symptoms. Subtropical and tropical commodities are especially sensitive to chilling, with chilling thresholds for tropical produce being around 13°C.

Manifestation of chilling injury is a function of time by temperature. Thus a short period at a certain temperature below the chilling threshold temperature may not result in the development of chilling injury symptoms in specific produce. However, relatively longer exposure will result in irreversible damage, the extent increasing with duration. Conversely, the development of chilling injury symptoms will be more severe in chilling-susceptible produce held for a shorter period well below the chilling threshold temperature.

### Adverse high temperature effects

Unusually high temperatures are associated with 'insults', such as exposure of harvested produce to direct sunlight, hot ambient air, and heat treatments for pest eradication (e.g. hot water dips, vapour and dry heat treatments). The activity of enzymes in fruit, vegetables and ornamentals declines at temperatures above 30°C. At certain temperatures, specific enzymes become inactive (denature); many are still active at 35°C, but most are inactivated at 40°C.

Continuous exposure of some climacteric fruit to temperatures around 30°C allows the flesh to ripen but inhibits fruit colouration. For example, the peel of Cavendish banana (cultivars Valery and Williams) remains green, and the accumulation of lycopene (red pigment) in tomato is inhibited during ripening at elevated temperature. When produce is held above 35°C, metabolism becomes abnormal and results in a breakdown of membrane integrity and structure, with disruption of cellular organisation and rapid deterioration of the produce. The changes are often characterised by a general loss of pigments, and the tissues may develop a watery or translucent appearance. This condition in banana and tomato is often referred to as 'boiled'.

## Pest activity interaction with temperature

Although, microbial and insect pest activities are considered in Chapters 9 and 11, respectively, it is important to note here that greatly diminished rates of pest population growth and development can be an additional benefit of good temperature management. For example, microbial growth is often only the visible symptom of wastage whose root cause is poor temperature management. Bacterial and fungal organisms exploit tissue deterioration (e.g. leakage of cellular contents) associated with adversely low or high temperatures, or simply with ongoing ripening and senescence processes in the normal physiological temperature range. Similarly, insect regeneration time can be decreased at low temperature. Accordingly, there is an obvious role for optimising temperature management to reduce reliance on unpopular chemical control measures.

## Sugar–starch balance

Storage of some vegetables — including potato, sweet potato, green peas and sweet corn — at low temperature can alter the starch–sugar balance in the produce. At any temperature, starch and sugar are in dynamic equilibrium, and some sugar is degraded to carbon dioxide during respiration:

starch $\leftrightarrow$ sugar $\rightarrow$ $CO_2$

At ambient temperatures, the starch—sugar balance in potato and sweet potato is heavily biased towards accumulation of starch. When these vegetables are stored at reduced temperature, the rate of respiration and the conversion of sugar to starch decreases. The critical temperature at which accumulation of sugar commences depends on the commodity (e.g. at about 10°C for potato and at about 15°C for sweet potato). Accumulation of sugar is profoundly undesirable in many starchy vegetables. Potato with a high sugar content has poor texture and a sweet taste when boiled; and when fried, excessive browning occurs due to caramelisation and reactions between amino acids and sugars (Maillard reaction). The accumulation of sugar in potato stored at low temperatures can be largely reversed by raising the storage temperature to 10°C or above. Although it is widely accepted that the sugar level returns to nearly normal during a week at 15–20°C, experience has shown that the decrease in sugar levels may occur at a much slower rate, especially after prolonged storage at low temperature.

In other vegetables, such as sweet corn and peas, a high sugar content is desired. These vegetables are harvested immature when the sugar content is highest, and rapid storage at quite low temperatures is necessary to retard the conversion of sugar to starch.

## Storage life

There is no one ideal temperature for the storage of all horticultural commodities because their responses to temperature vary widely. Important physical processes such as transpiration and physiological reactions like chilling injury must be taken into account, as must the required duration of storage. In fruits, vegetables and ornamentals not susceptible to chilling injury, maximum storage life can be obtained at temperatures close to the freezing point of the tissue. For chilling-sensitive produce, there are also potential immediate advantages of ensuring low temperatures during storage and distribution. These factors are considered in more detail in Chapter 13.

# COOLING OF PRODUCE
......

The object of cool storage is to slow deterioration without predisposing the commodity to abnormal ripening or other undesirable changes, thereby maintaining the produce in a condition acceptable to the consumer for as long as possible. Of those commodities that require low temperature storage, horticultural produce (compared, for example, with chilled meat or frozen chicken) are the most demanding for both the cool store operator and the engineer who designs the cool store and associated refrigeration equipment. In addition to providing sufficient refrigeration capacity to cool the produce to the required temperature, provision must be made for continuous removal of the heat of respiration (vital heat), maintenance of high RH, and, in some cases, ventilation or atmosphere control. Cool stores for fresh produce are generally required to operate within relatively close temperature limits (e.g. $\pm 1°C$), both in space (i.e. throughout the room) and time (i.e. constantly), in order to maximise storage life, avoid freezing, minimise desiccation and avoid gas injury (e.g. high carbon dioxide, low oxygen) of produce.

The temperature of horticultural produce at harvest is close to that of ambient air and can be in the order of $\geq 40°C$ for produce held in direct sunlight. At such temperatures respiration rates are extremely high, and storage life will be correspondingly reduced. It is often good practice to harvest early in the morning to take advantage of the lower produce temperatures generally prevailing at this time. Night harvesting has been investigated in certain situations, although the practice has not been sustained for reasons that include cost and inconvenience. Early morning harvesting may not be feasible for larger growers, and morning temperatures in tropical areas may still be relatively high.

It is axiomatic that the quicker the temperature of the produce is reduced to the optimum storage temperature, the longer will be its storage life. Rapid or fast cooling after harvest is generally referred to as 'precooling', and particularly benefits highly perishable (e.g. raspberries) and/or rapidly developing (e.g. asparagus) horticultural commodities. Special facilities for fast cooling of produce after harvest are often essential because refrigerated transport spaces — such as refrigerated ship holds, land vehicles and shipping containers — are generally not designed either to remove the field heat or to allow for adequate cold air circulation within and/or around individual containers. Rather, transport units are designed to maintain precooled produce at the selected carriage temperature. Rapid precooling was introduced in the early 1900s for long distance rail transport within the USA and for export fruit shipments. Maximum acceptable loading temperatures for perishable produce are now commonly and closely controlled in order to help ensure that produce will outturn, following refrigerated transport, in good condition.

The term 'precooling' is loosely applied, such that it encompasses any cooling treatment given to produce before shipment, storage or processing. A stricter definition of precooling would include only those methods by which the produce is cooled rapidly, certainly within 24 hours of harvest. No legal definition of precooling has been established. Thus the definition must be sufficiently broad and flexible as to embrace the cooling requirements of the various commodities in relation to their required postharvest life. The method of cooling selected will depend greatly on the anticipated storage life of the commodity. Rapidly respiring commodities, which have a short postharvest life, should be quickly cooled immediately after harvest. Commodities that have a longer postharvest life generally do not have to be cooled quite so rapidly, but nonetheless should still be cooled as soon as possible. Commodities that are susceptible to chilling injury should be cooled according to their individual requirements. Generic temperature recommendations for non-chilling-sensitive, moderately chilling-sensitive and highly chilling-sensitive commodity groupings are 0°C, 5–7.5°C and 13–15°C, respectively. Selection of the most appropriate precooling method depends on three main factors:

◆ the temperature of the produce at harvest;
◆ the physiology of the produce; and
◆ the desired postharvest life.

Produce that is to be 'cured' (Chapter 11) at temperatures above those required for extended storage is not normally precooled (e.g. potato, yam and sweet potato).

## Methods of cooling

Produce may be cooled by means of cold air (room cooling, forced-air or pressure cooling), cold water (hydrocooling), ice and evaporation of water (evaporative cooling, vacuum cooling). Fruit are normally cooled with cold air, although stone fruit benefit from hydrocooling. Any one of the cooling methods may be used for vegetables depending upon the structure and physiology of the specific commodities and upon market requirements/expectations. Cut flowers and foliage are usually forced-air cooled, although they may also be vacuum cooled if a small amount of water loss is acceptable.

## Cooling rates

The rate of cooling of produce depends primarily upon the:

◆ rate of heat transfer from the produce to the cooling medium, which is especially influenced by the rate of flow of the cooling medium around or through the containers of produce and the extent of contact between the two;
◆ difference in temperature between the produce and the cooling medium;
◆ nature of the cooling medium; and
◆ thermal conductivity of the produce.

When warm produce is exposed to cold air kept at a constant temperature by refrigeration, the rate of cooling (°C/h) is not constant, but diminishes exponentially as the temperature difference (driving force) between produce and air falls (Newton's Law) (Figure 4.3). Because the rate of cooling varies over time, two single parameters have been adopted to describe the cooling process. The cooling coefficient (C) is defined as the ratio of the change in product temperature per unit time (R; °C/h) at any moment to the difference in temperature between the product and the coolant (dT; °C) at the same moment:

$$C = R/dT$$

The 'half' and 'seven-eighths' cooling times are the times required to reduce the temperature difference between the product and the cooling medium by one half (Z) or by seven-eighths (S), respectively. Theoretically, Z and S (which is equivalent to three half-cooling times) are independent of the initial product temperature, and remain constant throughout the cooling period. S is of more practical use in commercial cooling operations because the temperature of the produce at the seven-eighths cooling time is acceptably close to the required storage or transport temperature. In systems where the cooling rate is rapid, the temperature

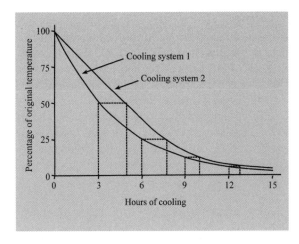

**Figure 4.3**
Cooling system 1 illustrates the ideal curve to achieve seven-eighths cooling (three half-cooling times, or 12.5% of the original temperature) in 9 hours. Cooling system 2 indicates the relative temperature change in the interior of a bulky commodity such as melons.

change in the interior of produce lags considerably behind the change in surface temperature. This is particularly true for bulky products, such as melons. In such cases, the limiting factor is the rate of heat conduction to the surface of the product. This positional lag effect can alter the relative difference in time between S and Z, such that S may range from 2Z to 3Z. Mathematically, seven-eighths cooling is expressed as:

$$S = \ln(8j)/C$$

where j is the lag factor, which may vary from 1 to 2 at the centre of the cooling objects, and C is the cooling coefficient, a negative value.

The cooling method, type of package, and the way the packages are stacked will all influence the rate of cooling of produce. The influence of such factors on the Z value are shown in Tables 4.2 and 4.3.

## Room cooling

Probably the most common precooling technique is room cooling, whereby produce in boxes (wooden, fibreboard, plastic), bulk containers or various other packages (e.g. mesh bags) is exposed to cold air in a normal cool store. For adequate cooling, air velocities around the packages should be at least 60 m/min. The design and operation of this system is simple (Chapter 7). Produce may be cooled and stored in the same place, thereby minimising rehandling. If produce is supplied over time, and as room cooling is relatively inefficient (i.e. slow), peak loads on the refrigeration system are less than those of faster cooling systems. However, in addition to slow cooling, room cooling has the disadvantage that comparatively more space is required. These disadvantages are especially pronounced when field bins or containers are unitised on pallets.

## Pressure (forced air) cooling

The rate of cooling with cold air may be significantly increased if the surface area available for heat transfer is enlarged by forcing air through packages and thus around each item of produce rather than only over the surfaces of the packages. Such 'pressure cooling' can cool produce in about 10–25 per cent the time required for room cooling.

Most commonly, pressure cooling involves passing cold air along an induced pressure difference (gradient) past initially warm produce in specially vented containers (Figure 4.4). The pressure differential is induced by fans that circulate cold air through the produce and packaging, which constitute the resistance to air flow. Pressure differentials between opposite faces of packages range from barely measurable to about 250 Pa (25 mm water head), and air flow rates can vary from 0.1–2.0 L/sec/kg. Within limits, the speed of cooling can be adjusted by varying the rate of air flow. Refrigeration requirements for pressure cooling are often overestimated because of a lack of understanding of the factors that limit the rate of heat loss from the cooling produce (cooling system 2, Figure 4.3). Such overestimation of refrigeration requirements increases the initial capital cost of a cooling plant unnecessarily. The thermal properties, S and j, of stacks of packaged produce should be taken into account when calculating refrigeration capacity.

### TABLE 4.2  HALF-COOLING TIMES FOR APPLES IN 18 KILOGRAM BOXES

| | Z(h) | |
|---|---|---|
| | APPLES LOOSE | APPLES WRAPPED |
| COOLING METHOD | IN BOX | AND PACKED |
| Conventional cool room | 12 | 22 |
| Tunnel, air velocity 200–400 m/min | 4 | 14 |
| High speed jet cooling, air velocity 740 m/min | 0.75 | |
| Hydrocooling (loose fruit) | 0.33 | |
| Single fruit air velocity 40 m/min | 1.25 | |
|  air velocity 400 m/min | 0.5 | |

SOURCE  E.G. Hall (1972) Precooling and container shipping of citrus fruits, CSIRO *Food Res. Quart.* 32: 1–10. (With permission.)

Maintaining high RH during pressure cooling is generally not considered to be critically important because the process is relatively rapid. However, this generalisation applies mostly to commodities with a low surface area to volume ratio, such as fruit and fruit vegetables. Indeed, high humidity pressure cooling systems (e.g. air-wash refrigeration systems) have been developed for commodities with a high surface area to volume ratio like leafy vegetables and cut flowers. Once a commodity has been adequately cooled, air velocity should be reduced or the produce transferred to a normal coolroom to reduce the risk of desiccation. When air-wash refrigeration systems are not in use, the risk of desiccation of horticultural produce during postcooling handling and storage is minimised by deliberately choosing conventional evaporator (refrigeration) coils with a high surface area, which allows the temperature difference between the refrigerant in the coils and the coldroom air passing over the coils to be kept smaller, thereby reducing condensation of water from the coldroom air onto the coils (frosting).

## TABLE 4.3 EXPECTED HALF-COOLING TIMES OF CENTRE FRUITS FOR VARIOUS PACKAGES AND STACKING PATTERNS

| | | Z(h) | |
| | FREELY | | |
| PACKAGE | EXPOSED | OPEN STOW★ | TIGHT STOW★ |
|---|---|---|---|
| 18 kg box loose fruit, unlidded | 7 | 18 | 45 |
| 18 kg box wrapped and packed | 23 | 35 | 45 |
| Cell pack 18 kg carton | 22 | 35 | 90 |
| Tray pack 18 kg carton | 23 | 43 | 90 |
| Half-tonne bin, lidded, no floor vents | 30 | 45 | 55 |
| Half-tonne bin, lidded, vents 8% floor area | 23 | 35 | 43 |
| Half-tonne bin, open-top | | | |
|     no floor vents | 18 | | |
|     5% floor vented | 11 | | |
|     10% floor vented | 5 | | |

★ Units of 44 boxes or cartons stacked on pallets.
SOURCE E.G. Hall (1972) Precooling and container shipping of citrus fruits, *CSIRO Food Res. Quart.* 32: 1–10. (With permission.)

**Figure 4.4**
Air flow in forced-air (pressure) cooling. Packages and containers must be vented (at least 5 per cent of exposed surface area) and stacked so that the cooling air is forced to flow through the containers and around the produce to return to the exhaust fans. A slight pressure drop occurs across the produce as air exhausts from the space between the stacks of produce.

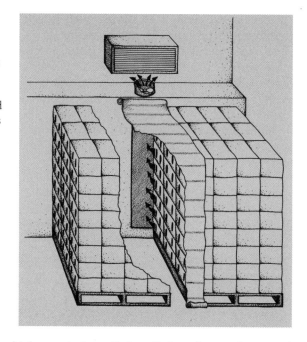

Condensation, which can induce fruit splitting, favour decay and weaken paper-based packaging materials, is not a problem with pressure cooling. Cold air will progressively warm and increase its water holding capacity as it passes through the package. In contrast, condensation may be a problem with room cooling, where transpired water vapour may condense on the faster cooling outermost packaging and produce (Chapter 5).

## Hydrocooling

In hydrocooling, water is the heat transfer medium. Thus, both produce and containers must be tolerant to wetting. Since water has a far greater heat capacity than air, hydrocooling is comparatively rapid provided that the water contacts most of the surface of the produce and is maintained close to the prescribed temperature, usually 0°C. In many hydrocooling systems, produce passes under cold showers on a continuous feed conveyor. Alternatively, cold water baths and/or batch process systems may be used. Hydrocooling may also help clean the produce. However, greater contamination of produce with spoilage microorganisms can occur if soil and debris are not removed from the system (e.g. allowed to settle, or filtered) and the water renewed and/or disinfected. A further advantage of hydrocooling is that the commodity loses little weight during hydrocooling. When cooling is completed, produce must be moved to a coolroom to prevent rewarming.

## Icing

Before the advent of comparatively modern precooling techniques (e.g. pressure cooling), contact or package icing was used extensively for precooling produce and maintaining temperature during transit, particularly for the more perishable commodities such as leafy vegetables. The melting of ice to water (latent heat of fusion) absorbs 325 kJ (heat energy) per kg. Contact icing is now mainly employed as a supplement to other forms of precooling. In top icing, finely crushed ice or an ice slurry (liquid ice comprising approximately 40 per cent water, 60 per cent ice and 0.1 per cent salt) is sprayed onto the top of the load of produce, generally inside the road or rail transit vehicle. In some distribution and marketing systems it is common practice — but largely unnecessary, expensive, and also undesirable due to the promotion of soft rots — to add crushed ice to precooled produce such as broccoli and sweetcorn packed in polystyrene containers. The alleged intention is to maintain freshness during transport and marketing. However, for the reasons mentioned above, the practice of package icing should be discouraged in favour of refrigerated handling and transport. Nevertheless, it is a widely used method in transportation by air freight because of the difficulties of maintaining temperature control throughout the complex handling regime at airports.

## Vacuum cooling

Vegetables such as lettuce, with a high surface to volume ratio, may be quickly and uniformly cooled by boiling off some of the constituent water at reduced pressure. The rate of cooling is at least as rapid as that achieved with hydrocooling. In principle, vacuum cooling is evaporative cooling, exploiting the latent heat of vaporisation of water. Produce is loaded into a sealed container, and the pressure is reduced to about 660 Pa (5 mm mercury) (Figure 4.5). At this sub–atmospheric pressure (cf. atmospheric pressure of 760 mm mercury), water boils at 1°C. For every 5°C drop in temperature, approximately 1 per cent of the produce weight is lost as water vapour. This water loss may be minimised by spraying the produce with water, either before enclosing it in the vacuum chamber or towards the end of the vacuum cooling operation (hydro-vacuum cooling). The rate of vacuum cooling is largely dependent on the surface to volume ratio of the produce and on the ease with which the commodity loses water. Leafy vegetables are ideally suited to vacuum cooling. Other vegetables, such as asparagus, broccoli,

**Figure 4.5**
Vacuum cooler designed for cooling two pallets of fresh produce at a time. (Courtesy of D. Harding P/L, Highbury, South Australia.)

Brussels sprout, mushroom and celery, may also be successfully vacuum cooled, as can cut flowers and foliage. Fruit, which has a low surface to volume ratio and a waxy cuticle and which therefore lose water slowly, do not benefit by vacuum cooling. Comparative cooling of several vegetables are shown in Table 4.4.

### TABLE 4.4 COMPARATIVE COOLING OF VEGETABLES UNDER SIMILAR VACUUM CONDITIONS

| PRODUCE | INITIAL TEMPERATURE (20°C) FINAL TEMPERATURE (°C) |
|---------|---------------------------------------------------|
| Lettuce, onion | 2 |
| Sweet corn | 5 |
| Broccoli | 6 |
| Asparagus, cabbage, celery, pea | 7 |
| Carrot | 14 |
| Potato, zucchini | 18 |

SOURCE Adapted from American Society of Heating, Refrigerating and Air-conditioning Engineers (1986) *ASHRAE handbook of refrigeration systems and applications*, Atlanta, GA.

### Evaporative cooling

This is a simple process in which dry air is cooled by blowing it across a wet surface. The evaporation of water (latent heat of vaporisation) absorbs 2260 kJ (heat energy) per kg. The technique is efficient only under conditions of low ambient RH (e.g. summer in Mediterranean-type climates; arid and semi-arid regions), and requires a good quality water supply. The low energy cost of evaporative coolong is an advantage. The commodity may be cooled either by the humidified cool air, or by misting produce with water and then blowing dry air over the wet commodity. The extent to which air may be cooled by the evaporation of water is a function of the water holding capacity of the air, which in turn is a function of temperature and relative humidity (see Chapter 5). Evaporative cooling might be considered suitable for citrus, which do not require very low temperatures and are grown in dry environments such as Southern California, Israel, and South Australia.

## Night air cooling

Night air cooling involves circulating relatively cool 'night air' around produce kept within an insulated store. The cooler night air can be used to remove vital heat (i.e. heat of respiration) and other forms of heat (e.g. conducted heat) that accumulates in the store during the day. This cooling method, which may be considered a variation on room cooling, is useful only for chilling sensitive produce, such as citrus fruit. Even then, it is only applicable in areas where large day/night (diurnal) variations in temperature prevail (e.g. semi-arid and arid regions).

# FURTHER READING

......

American Society of Heating, Refrigerating and Air-conditioning Engineers (1986) *ASHRAE handbook of refrigeration systems and applications*, Atlanta, GA.

Hallowell, E.R. (1980) *Cold and freezer storage manual* (2nd ed.), AVI, Westport, CT.

Hardenburg, R.E., A.E. Watada and C.Y. Wang (1986) *The commercial storage of fruits, vegetables and florist and nursery stocks* (rev. ed.), Agriculture Handbook, no. 66, US Department of Agriculture, Washington, DC.

Kays, S.J. (1991) *Postharvest physiology of perishable plant material*, Van Nostrand Reinhold, New York.

Kitinoja, L. and A.A. Kader (1994) *Small-scale postharvest handling practices: A manual for horticultural crops* (2nd ed.), Department of Pomology, UC Davis, CA.

Lipton, W.J. and J.M. Harvey (1977) *Compatibility of fruits and vegetables during transport in mixed loads*, US Department of Agriculture, Marketing Research, Washington, DC. Report no. 1070.

Ryall, A.L. and W.J. Lipton (1979) Handling, transportation and storage of fruits and vegetables (rev. ed.), vol. 1, *Vegetables and melons*, AVI, Westport, CT.

Ryall, A.L. and W.T. Pentzer (1982) Handling, transportation and storage of fruits and vegetables (rev. ed.), vol. 2, *Fruits and tree nuts*, AVI, Westport, CT.

Story, A. and D.H. Simons (eds) (1997) *Fresh produce manual*, Australian United Fresh Fruit and Vegetable Association Ltd, Sydney.

Thompson, A.K. (1996) *Postharvest technology of fruit and vegetables*, Blackwell Science, Oxford.

Wade, N.L. (1984) Estimation of the refrigeration capacity required to cool horticultural produce, *Int. J. Refrigeration*, 7: 358–66.

Watkins, J.B. (1990) *Forced-air cooling*, Queensland Department of Primary Industries, Brisbane. Information series UQ188027.

Whiteman, T.M. (1957) *Freezing points of fruits, vegetables and florist stocks*, US Department of Agriculture and Marketing, Washington, DC. Research Report no. 196.

# 5
# WATER LOSS AND HUMIDITY

Fresh fruit, vegetables and ornamentals are mostly composed of water, the unique 'universal solvent' that is fundamentally important in all life processes. Accordingly, horticultural commodities might be regarded as water in aesthetically pleasing packages. As water is rather costly to put into these attractive packages, it is not surprising that the final product is relatively expensive. It follows, then, that water loss equates to loss of saleable weight, and thus constitutes a direct loss in marketing. Accordingly, measures that minimise water loss after harvest will usually enhance profitability. Loss in weight of only 5 per cent will cause many perishable commodities, even bulky fruit with a low surface area to volume ratio, to appear wilted or shrivelled. Under warm, dry conditions this level of water loss can occur from some produce in just a few hours. It almost goes

**Figure 5.1**
Relationship between weight loss and green life of mango fruit at 20°C. Green life was measured as the time for the fruit to yield 0.55 mm using a modified tomato compression meter. The equation of the line is $y = 13 - 1213x$ ($r^2 = 0.63$, $n = 50$).
SOURCE Adapted from A.J. Macnish, D.C. Joyce and S.E. Hetherington (1997, in press) Packaging to reduce water loss can delay ripening of mango (*Mangifera indica* L. cv 'Kensington Pride') fruit. *Aust. J. Exp. Agric.* 37: 463–67.

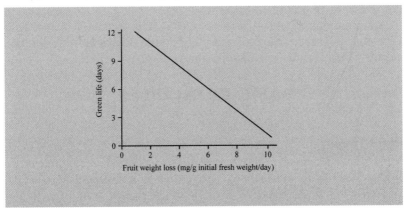

**Figure 5.2**
Loss of ascorbic acid (vitamin C) during storage of exposed (open) and wrapped in plastic (closed) *Brassica juncea* leaves at 24–28°C. Initial content of vitamin C was about 18 µg/cm². Adapted from H. Lazan, Z.M. Ali and F. Nahar (1987) Water stress and quality decline during storage of tropical leafy vegetables, *J. Food Sci.* 52: 1286–88.

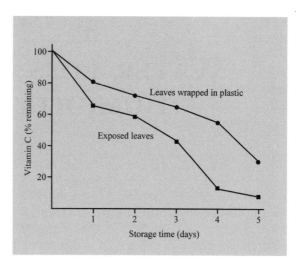

without saying that wilted cut flowers and foliage and pot plants have little or no consumer appeal. Even in the absence of visible wilting, water loss can result in reduced crispness and/or early ripening of some fruit (Figure 5.1), and undesirable changes in colour, palatability and nutritional quality may ensue in some vegetables (Figure 5.2).

In contrast to conditions that promote water loss, conditions that result in wetting of produce can also result in disastrous losses with some commodities. Free water encourages microbial decay and, in some cases, such as grapes and apricots, causes physical splitting of commodities. Additionally, free water can encourage undesirable growth, such as: callus growth from lenticels of mature green avocado fruit or fresh green beans held in a water-saturated atmosphere; browning, callusing and rooting on cut surfaces of broccoli stems under similar conditions; and rooting and sprouting in onions and potato.

In examining the role of water in quality maintenance, it is necessary to both understand the basic principles involved and to recognise specific physiological responses of different commodities

# BASIC PRINCIPLES
·······

### *Plant tissue*

A steady water supply is essential to the physical development of produce on a plant, but most harvested horticultural produce — except for well watered pot plants, or cut flowers and foliage stood in water —

is removed from such a water supply. The physical development of produce on the plant is at least partly made possible by positive turgor driving cell expansion. Turgor is also a principal means of support in herbaceous material. Turgor potential ($\Psi_{TP}$) is one of the two major components of water potential ($-\Psi_{WP}$). The other major component is osmotic potential ($-\Psi_{OP}$), which is negative relative to pure water ($\Psi = 0$) because of the dissolved organic (e.g. sugars) and inorganic (e.g. salts) solutes that lower the free energy of water due to hydrogen bonding. These three parameters are related diagrammatically in Figure 5.3 and by the following equation:

$$-\Psi_{WP} = \Psi_{TP} + (-\Psi_{OP})$$

$\Psi_{OP}$ can be related to the concentration of individual osmotica by the following relationship (Van't Hoff equation):

$$\Psi_{OP} = -m \cdot i \cdot R \cdot T$$

where, for each individual solute, m is the molal concentration, i is the ionisation constant (e.g. 1 for a non–dissociating solute such as glucose, 2 for a salt that dissociates into two ions (such as NaCl), R is the universal gas constant (0.00831 L.MPa/mol/°K) and T is the absolute temperature (°K = °C+273.15). $\Psi_{OP}$ values calculated using the Van't

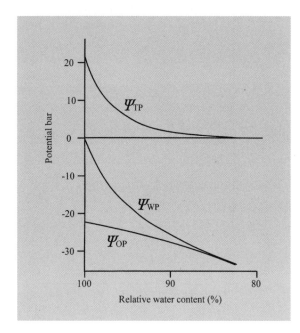

**Figure 5.3**
A Hofler diagram derived from analysis of the $Y_{WP}$ isotherm and showing the relation between relative water content (RWC) and water potential ($\Psi_{TP}$), osmotic potential ($\Psi_{OP}$) and turgor potential ($\Psi_{TP}$).
SOURCE Adapted from M.T. Tyree and P.G. Jarvis (1982) Water in tissues and cells, in O.L. Lange, P.S. Nobel, C.B. Osmond and H. Ziegler, *Physiological plant ecology*, vol. II, *Water relations and carbon assimilation*, Springer-Verlag, Berlin.

Hoff relationship on the basis of published compositional data (i.e. the sum of the $\Psi_{OP}$ values calculated for each solute) for ripe mango and avocado fruit are -1.7 and -0.6 MPa, respectively. These calculated values agree reasonably well with measured values.

Relative to water in the xylem (i.e. the transpiration stream), the osmotic potential of water in plant cells is low. Thus, water moves from the xylem into cells along (down) a gradient in water potential. Water will accumulate in cells as a result of this gradient, and therefore exert pressure against the restraining cell walls. While this positive turgor pressure acts to counter the negative cell osmotic potential, the net negative water potential that results is sufficiently low to continue to draw water from the xylem.

In intact plants, the xylem water is often under tension due to transpirational pull. Should the xylem water potential ($\Psi_X$) fall below the water potential of adjacent cells, water may be withdrawn from these cells. For example, fruit often shrink (i.e. decrease in volume) during the course of a clear sunny day as a consequence of high transpirational demand imposed by nearby leaves (Figure 5.4). This increased transpirational demand is the result of an increasing vapour pressure deficit, which is associated with increasing temperature and falling RH.

**Figure 5.4**
Change in the radius of attached mango fruit in relation to vapour pressure deficit (VPD, kPa) over a 3-day period.
SOURCE D. Beasley and D. Joyce (unpublished data).

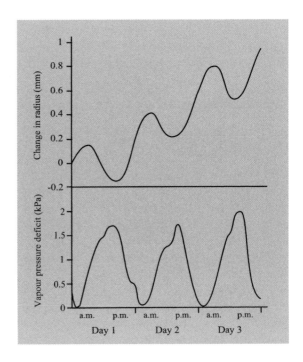

After harvest, continuing transpiration in the absence of a supply of water may soon dehydrate plant tissue since the water potential of warm and relatively dry air is much lower than that of plant tissue. Initially, shrinkage will occur, followed by wilting when turgor falls to zero (Figure 5.3). However, wilting may not be obvious in all tissues, especially those with a high degree of mechanical support associated with thick walled cells (e.g. collenchyma cells) and/or vascular tissue (e.g. water-conducting xylem vessels). The degree to which tissue can lose water prior to wilting (i.e. the water storage capacity) depends on the initial level of hydration and on the elasticity of the walls of the cells comprising the tissue. Elastic cell walls stretch under the pressure of turgor. Thus, elastic cells can lose a proportionally greater volume of water, and contract around the shrinking protoplast, before they reach zero turgor. Thereafter, dehydration and desiccation will be apparent. Along with water loss there is a concomitant decrease in osmotic potential (Figure 5.3).

The above scenario would be invariably and rapidly enacted in harvested horticultural produce were it not for the ingenious natural 'packaging' of plant material. This natural packaging involves structural barriers, such as the waxy cuticle (in essence, a 'plastic bag') and physiological controls, such as stomatal closure (in essence, 'variable aperture microperforations'). This packaging confers a high resistance to water loss in most crops.

The solutes in plant tissues slightly depress the vapour pressure of the water. Thus, when fresh plant tissue is placed within a water barrier under absolutely isothermal conditions, the air will not become completely saturated due to the presence of these solutes (-$\Psi_{OP}$) and, to a much lesser extent, bound water (matrix potential; $\Psi_M$). An equilibrium relative humidity (ERH) of at least 97 per cent is associated with most plant tissues. ERH can be related to water (or osmotic) potential by the following equation:

$$\text{ERH } (\%) = e^{[-\Psi_{vw} (V_w / R \cdot T)]} \times 100$$

where $\Psi_{WP}$ is the water potential of water vapour, $V_W$ is the molar volume of water (e.g. 18.048 x $10^{-6}$ m$^3$/mol at 20°C), R is the gas constant and T is the absolute temperature.

Measures of ERH are of particular interest to microbiologists because small reductions in water activity ($a_W$, where $a_W = \text{ERH}/100$) and water potential equivalent to a reduction in ERH by about 95 per cent can inhibit the growth of most bacteria and fungi in culture (Figure 5.5). However, growth of some pathogenic fungi is not prevented until water activity and water potential are reduced to less than the equivalent of about 85 per cent ERH.

**Figure 5.5**
Growth in colony diameter of *Botrytis cinerea* on a solid medium after 3 days at various equilibrium relative humidities (ERH) at 20°C. The ERH were adjusted with sucrose. SOURCE Adapted from S. Alam, D. Joyce and A. Wearing (1996) Effects of equilibrium relative humidity on *in vitro* growth of *Botrytis cinerea* and *Aternaria alternata*. *Aust. J. Exp. Agric.* 36: 383–88.

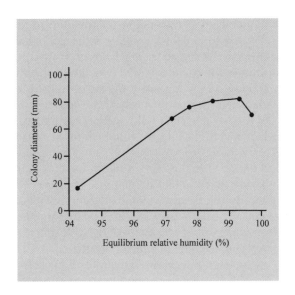

## Vapour pressure deficit

Moist air consists of a mixture of dry air and water vapour. 'Humidity' is the general term referring to the presence of water vapour in air. If pure water is placed in an enclosure containing dry air, water molecules will enter the vapour phase until the air becomes saturated with water vapour. The amount of water vapour in atmospheric air can vary from zero to a maximum that is dependent on temperature and pressure. For example, saturated air at 30°C contains approximately 4 g water vapour /100 mL of air.

'Relative humidity' (RH) is probably the best known term for expressing the humidity of moist air. RH is defined as the ratio of the water vapour pressure in the air to the saturation vapour pressure possible at the same temperature, expressed as a percentage. Saturated air, therefore, has an RH of 100 per cent. When water-containing material, such as fruit, is placed in an enclosure filled with air, the water content of the air increases or decreases until the ERH is reached; that is, when the number of water molecules entering and leaving the vapour phase is equal. ERH is a property of the material and its moisture content. Pure water has an ERH of 100 per cent.

The water potential of air can be related to RH by the following relationship:

$$-\Psi_{wv} = (RT / V_w) \cdot \ln(RH\%/100)$$

which is the equation for ERH(%) rearranged (see above). The water potential of air is usually low relative to that of harvested horticultural produce (e.g. -14.2 MPa at 90 per cent RH and -93.6 MPa at 50 per cent RH; 20°C).

Psychrometric charts that graphically relate the various properties of moist air (e.g. wet bulb and dry bulb temperatures, water vapour pressure) have been constructed (Figure 5.6). The scale along the bottom axis of Figure 5.6 indicates dry bulb temperatures as given by a wet-and-dry bulb hygrometer. Dry air at all temperatures has zero water concentration, and therefore zero water vapour pressure. The curved line at the top of Figure 5.6 illustrates the relationship between vapour pressure and temperature in saturated air. By definition, this is the line of 100 per cent RH. Other curved lines can be drawn representing constant RH over a range of temperatures. The curvature of these lines shows that the vapour pressure (vp) of water increases rapidly with temperature. For example, at 30°C in saturated air (100 per cent RH) the vp of water is 4.3 kPa; at 20°C the vp is less at 2.4 kPa; and at 10°C it is only 1.3 kPa.

The difference between the vp of the produce (which is a function of temperature and ERH) and that of the surrounding air (which is a function of temperature and RH) is called the 'vapour pressure deficit' (vpd). The vpd has important consequences in the cooling of fresh produce. Even if saturated cool air is used to cool produce, as long as the produce remains warmer than the cooling air it will lose water. Consider, for example, produce at 20°C and 97 per cent ERH and humidified air at 0°C and 100 per cent RH. The vp of the produce will be about 2.4 kPa, whereas that of the air will be about 0.6 kPa (Figure 5.6). Thus, the vpd will be:

2.4 kPa (produce) — 0.6 kPa (air) = 1.8 kPa

This gradient is effectively the magnitude of the concentration

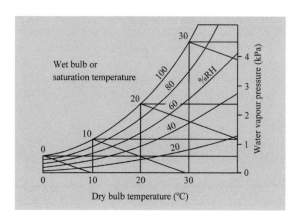

**Figure 5.6**
Simplified psychrometric chart.

gradient driving water loss (i.e. $F = R \cdot vpd$), where F is the flux of water molecules and R is a constant whose magnitude is determined by produce characteristics such as cuticular properties. Thus it is important to cool produce rapidly to minimise the vpd between the produce and the cooling air stream.

Calculations of vpd can be applied to compare potential relative rates of water loss under different sets of conditions. In the above example, produce at 20°C and 97 per cent ERH was placed in a 0°C airstream at 100 per cent RH (condition 1). The resultant vpd was about 1.8 kPa. Consider now the relative rate of water loss for produce already cool at 5°C (97 per cent ERH) and being placed into the same airstream (0°C, 100 per cent RH) (condition 2). The vp of this produce will be 0.9 kPa (5°C, 97 per cent RH) and the vp of the airstream will still be 0.6 kPa. Thus the vpd for this second set of conditions will be 0.3 kPa (i.e. 0.9 kPa — 0.6 kPa). Thus initial water loss under condition 1 is likely to be sixfold that under condition 2.

Another important physical property of moist air is evident in Figure 5.6, namely dew point. When moist air is cooled, a temperature is reached at which the vp of water reaches the maximum for that temperature. Thereafter, with further cooling, water will condense. For example, fog is formed on a cooled surface when water condenses. The temperature at which condensation occurs is the dew point temperature. The horizontal lines in Figure 5.6 can be used to find dewpoint temperatures. Thus, air of 80 per cent RH at 30°C becomes saturated when cooled to about 26°C. The dew point temperature is equal to the dry bulb temperature at the point of intersection with the saturation curve. The lines that slope upwards from right to left indicate constant wet bulb temperatures.

As mentioned above, condensation forms on cold surfaces. For instance, if a fibreboard carton of warm transpiring produce is placed in a cold room, the carton will cool relatively quickly and water will condense onto it. The result can be catastrophic in terms of physical damage to produce, since the strength of moist or wet fibreboard is substantially reduced and may cause the carton to collapse. In an alternative scenario, condensation will form on cooled produce in warm moist air (e.g. a warming carton). This condensation can promote rotting through spore germination, accelerate warming of the produce (through the release of the latent heat of vaporisation under conditions of high thermal conductivity), induce splitting through increased turgor, and interfere with respiratory and other gas exchange (e.g. ethylene) as the diffusion of gas in the condensed water is several orders of magnitude lower than in air.

Under conditions of low storage temperatures (e.g. 0°C) and high RH (e.g. 95 per cent), extremely small fluctuations in temperature (<0.5°C) can result in condensation on cooling surfaces. For example, Figure 5.7 presents a graphical relationship between RH and water potential, $\Psi_{WP}$. Superimposed on this graph are data on sugar (glucose) concentration and dewpoint. From Figure 5.7 it can be seen that if a produce, such as grape with 15 per cent sugar being the main osmoticum, is sealed in a waterproof bag (e.g. polyethylene liner), the ERH and water potential of the atmosphere in the bag will be about 97 per cent and -3.3 MPa, respectively. Under such conditions, condensation (dew formation) will result if the temperature of the plastic bag falls by just 0.38°C (Figure 5.7).

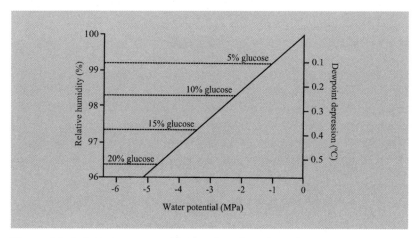

**Figure 5.7** Relation between equilibrium relative humidity and water potential of plant tissues at 0°C. The inserted scale relates the temperature of the dew point to different values of water potential and the concentrations of glucose that give these values of water potential in the absence of turgor (osmotic potential).
SOURCE Adapted from B.D. Patterson, J. Jobling and S. Moradi (1994) Water relations after harvest—new technology helps translate theory into practice. Australasian Postharvest Conference, The University of Queensland, Gatton College, 1993, pp. 99–102.

# FACTORS AFFECTING WATER LOSS
·······

As a function of their physicochemical characteristics, horticultural produce vary widely in their inherent rates of postharvest water loss. Some examples of relative rates of water loss, presented as transpiration coefficients (mg water/kg product/sec/MPa vpd), are given in Table 5.1.

## TABLE 5.1 TRANSPIRATION COEFFICIENTS FOR SELECTED HORTICULTURAL PRODUCE

| PRODUCE | TRANSPIRATION COEFFICIENT (MG/KG.SEC/MPA) | RANGE (FROM LITERATURE) |
|---|---|---|
| Apple | 42 | 16–100 |
| Brussels sprout | 6150 | 3250–9770 |
| Cabbage | 223 | 40–667 |
| Carrots | 1207 | 106–3250 |
| Grapefruit | 81 | 29–167 |
| Leek | 790 | 540–1042 |
| Lettuce | 7400 | 680–8750 |
| Onion | 60 | 13–123 |
| Parsnip | 1930 | 1094–2771 |
| Peach | 572 | 142–2089 |
| Potato | 25 | 15–40 |
| Tomato | 140 | 71–365 |

SOURCE S. Ben-Yehoshua (1987) Transpiration, water stress and gas exchange, in J. Weichmann (ed.), *Postharvest physiology of vegetables*, Marcel Dekker, New York.

## Produce factors

### Surface area/volume ratio

A major factor in the rate of water loss from produce is its surface area to volume ratio. On purely physical grounds there is proportionally greater loss by evaporation from produce with a high surface area to unit volume ratio. For instance, it is estimated that individual edible leaves have surface area (SA) to volume (V) ratios of around $50–100$ $cm^2.cm^{-3}$, whereas tubers have SA:V ratios of about $0.5–1.5$ $cm^2/cm^{-3}$. Thus, all other factors being equal, a leafy vegetable (e.g. spinach) or a cut flower (e.g. rose) will lose moisture, and therefore weight, much faster than a fruit (e.g. apple). Similarly, a small fruit, root or tuber will lose weight faster than a relatively large one.

*Plant surfaces*

The nature of plant surfaces and the underlying tissues of fruit and vegetables have a very marked effect on the rate of water loss. Many types of produce have a waxy coating on the surface known as the cuticle, which is highly resistant to the passage of water or water vapour. Thus by restricting water loss through evaporation, the cuticle plays an essential role in maintaining the high water content within tissues that is necessary for normal metabolism and growth. It has been estimated, for example, that the cuticle reduces the rate of evaporation from living plant cells from about 3.6 to 0.14 $mg/cm^{-2}.hPa^{-1}.h^{-1}$.

The structure of the cuticle may be more important than its thickness. Cuticles that consist of a complex, well ordered structure of overlapping platelets provide greater resistance to the permeation of water vapour than coatings that are thicker but flat and structureless. In the former case, the water vapour must follow a more diverse path for it to escape into the atmosphere. This tortuous pathway and the thickness of the unstirred boundary layer over the surface of the harvested organ may be further extended by trichomes (hairs). Trichomes vary widely in shape (e.g. club or branched) and structure (e.g. single or multicelled). These hairs originate from the epidermis, which underlies the waxy cuticle. Epidermal cells are compactly structured, with minimal space between adjacent cells.

The bulk movement of water vapour and other gases (viz. oxygen, carbon dioxide) in and out of leaves is controlled by small pores called stomata, or stomates (Figure 5.8). These pores are located at intervals in the epidermis. Below the stomata are substomatal cavities that connect

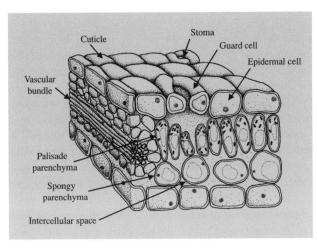

**Figure 5.8**
Cross-section of a leaf showing the surface (cuticle), stoma, internal structures and the network of intercellular spaces.

with the intercellular airspace network. The stomata in harvested leafy produce normally close when the paired guard cells on either side lose turgor in response to a small amount of water loss. However, under some conditions, such as rapid cooling of chilling-sensitive tissues, the stomata may remain open.

Many fruit (e.g. apple) and storage organs (e.g. sweet potato) have lenticels, and not stomata, on their surface. Lenticels are sunken openings that are somewhat similar to stomata but that are larger and generally contain tightly packed and suberised hypodermal cells. Lenticels may derive initially from stomata that, because of their rigid guard cell structure, were torn apart, such as during rapid fruit development, for example. Suberin, like the cutin of the cuticle, is hydrophobic, and therefore serves to minimise water loss. Otherwise, there is no mechanism for 'closing' lenticels. In mature fruit, in particular, lenticels are often blocked with wax and debris. Thus, loss of water as vapour, and also respiratory gas exchange, may only take place by diffusion through the cuticle. In some harvested fruit (e.g. tomato), a large stem scar can also be an important route for water loss.

The surface of some tubers and roots consists of a periderm (cork cells), which comprises several layers of suberised cells. The periderm is generated from a layer of cambial cells. Periderm formation is necessary when underground organs (e.g. potato tubers) are exposed to the atmosphere's relatively low humidity. In the case of sweet potato, incubation conditions of high humidity (90–95 per cent RH) and elevated temperature (29°C) (termed 'curing') are utilised after harvest for about 1 week to encourage periderm formation.

## Mechanical injury

Mechanical damage can greatly accelerate the rate of water loss from produce. Bruising and abrasion damages the surface organisation of the tissue, thereby allowing much greater flux of water vapour through the damaged area. Cuts are of even greater importance as they completely break the protective surface layer and directly expose underlying tissues to the atmosphere. If the damage occurs early in the growth and development of produce, the organ usually seals off the affected area with a layer of corky callus cells. The capacity for wound healing generally diminishes as plant organs mature, so damage that occurs during harvesting or in postharvest operations generally remains unprotected. However, some mature produce, notably tubers and roots, retain the capacity to seal off wound areas. Again, curing at suitable temperatures and humidities encourages wound healing (see Chapter 11). Damage to surface tissue can also occur as a consequence of attack by pests (insects, rodents) or diseases, and will result in increased rates of water loss.

If high concentrations of solutes are released upon wounding (e.g. sugars from wounded grapes), their osmotic property can attract water vapour, and a droplet may form. Such droplets may continue to grow in volume under conditions of high in–package RH and continued extraction of osmotica from the tissue. Droplets provide free moisture and nutrients to support fungal spore germination and sustain mycelial growth and sporulation.

## CONTROL OF WATER LOSS

Plugging lenticels with applied wax (e.g. apples) or harvesting fruit at a more advanced maturity stage (e.g. marrows versus zucchini) might be considered a means of modifying or utilising, respectively, tissue structure to limit water loss. However, there is only limited scope for reducing rates of water loss by such means. The most important methods of reducing the rate of water loss from produce primarily involve lowering the capacity of the surrounding air to hold additional water. This objective is achieved by lowering the temperature and/or raising the RH (i.e. by reducing the vpd between the produce and air). The effect of vpd on water loss from apple cultivars is shown in Figure 5.9. An alternative to raising RH is to provide a barrier to water loss (e.g. wax or other hydrophobic coatings [Chapter 11] or plastic film [Chapter 6]).

**Figure 5.9**
Effect of vapour pressure difference (deficit) on weight loss by the apple cultivars Golden Delicious (upper curve) and Jonathan (lower curve).
SOURCE Adapted from A.W. Wells (1962) *Effects of storage temperature and humidity on loss of weight by fruit*, US Department of Agriculture, Washington DC, Marketing Research Report no. 539.

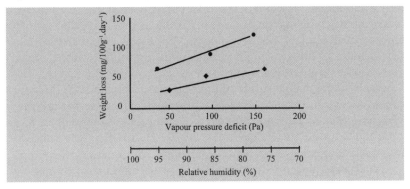

## *Increasing humidity*

Increasing the RH of the air reduces the vpd between the produce and the air, and hence the amount of water evaporated from the produce before the surrounding air is saturated. However, the use of very high RH (e.g. >95 per cent) generally favours the growth of moulds (e.g. on citrus). Thus, at very high humidities, the potential benefits of reducing water loss by plant tissues generally tend to be outweighed by the risk of rotting. In culture, most fungi cease to grow when the RH is reduced to about 90 per cent and only a few can grow at 85 per cent. Under drier conditions spores cannot germinate and, even if there is enough free moisture in a wound to permit germination, a dry atmosphere may dry the exposed tissue fast enough to prevent infection and development of a rot. Experience has shown that an RH of 90 per cent is usually the best compromise condition for the storage of fruit. However, an RH of 98–100 per cent remains the best condition for leafy vegetables, some root vegetables, and for cut flowers and foliage, which have relatively high coefficients of transpiration. Fungicides may be employed to minimise or overcome the problem of fungal growth at high RH. However, consumer resistance to the use of such chemicals should be considered. Storage of potato continuously at high humidity has an added advantage in that it is less prone to develop pressure bruises than if stored at low humidity. Nevertheless, commodities such as potato and onion are also more likely to sprout at high RH. In contrast to most harvested horticulture produce, onion and mature cucurbits (e.g. pumpkins) require a lower in-storage RH of 65–70 per cent in order to prevent excessive rotting.

It is a relatively simple procedure to increase the RH of air. This can be achieved by spraying water as a fine mist, by introducing steam, or by increasing the temperature of the refrigeration coils. The addition of water vapour to a cold storage chamber can be controlled automatically with a humidistat. However, a potential problem of adding free water to a system is its condensation on cold surfaces (e.g. produce, walls, cartons) and eventual pooling (e.g. on floors). An associated problem is the development of frost on cold refrigeration coils, necessitating longer and/or more frequent defrosting cycles.

For produce in cold rooms, the best way to maintain high RH (e.g. 95 per cent) is to maintain a small temperature difference between the refrigeration coils and the produce by using large coils with a high surface area for heat exchange (see Chapter 7). Another approach is to use an air wash refrigeration system (Chapter 7).

Fruit ripen better — with not only better appearance due to the absence of shrivelling but also with better internal quality — at an RH of at least 90 per cent. For instance, the necessity for controlling humidity in banana ripening rooms is generally well recognised. In contrast, storage of some apple varieties at very high humidities can be a disadvantage. This is because a relative increase in weight loss from cultivars that are susceptible to internal or low temperature breakdown can actually decrease the incidence of the disorder.

### Air movement

Air movement over the produce is a highly significant factor influencing the rate of moisture loss. While air movement is required to remove heat from produce, its effect on moisture loss must also be considered. There is always a thin unstirred layer of air adjacent to the surface of the produce. In this layer, the water vapour pressure is approximately in equilibrium with that of the produce itself. Air movement tends to sweep away this moist air from around the produce. Increasing the rate of air movement reduces the thickness of the boundary layer and increases the vapour pressure difference near the surface, and so increases the rate of moisture loss. Thus, in a cool store, restricting the air movement around the produce can effectively reduce the rate of water loss.

After initial cooling, reduced air movement can be achieved by running the forced draft cooler fans at a lower speed or by reducing the length of time that they are operated (Chapter 7). Open rooms with natural ventilation can be modified to restrict the flow of air. It should be recognised that regulation of air movement requires a compromise. Sufficient air movement is needed to prevent large temperature gradients being produced within the storage chamber but at rates that will minimise water loss from the produce. An 'extreme' form of cool storage that maintains high RH by eliminating both a vpd and forced air movement is the jacketed cool store. In this system, cold air is circulated around a plastic tent that contains the produce and is erected within a cool store. On a smaller scale, the same condition is achieved with plastic carton liners.

### Air pressure

Reduced air pressure (e.g. during vacuum cooling or during air freight under partial pressurisation conditions) will increase the rate of water loss. Thus the time produce spends under reduced pressure conditions must be minimised and/or measures taken to inhibit water loss, such as by misting with water or providing moisture barrier packaging, respectively.

## Packaging

Water loss can be very effectively reduced by placing an additional physical barrier around the produce, which also reduces air movement across its surface. Simple methods are to pack the produce into bags, boxes or cartons and to cover stacks of produce with tarpaulins. Close packing of produce itself restricts the passage of air around individual items, and thus reduces water loss. Hence, placing produce in mesh bags can have a beneficial effect because of the closer packing. That is, thick unstirred boundary layers are associated with the inner produce items, and these are consequently protected from direct exposure to dry air by the outer layers in the bags.

The degree to which the rate of water loss is reduced by packaging depends on the permeability of the package to water vapour transfer as well as on the closeness of containment. All commonly used materials are permeable to water vapour to some extent. Materials such as polyethylene film are excellent vapour barriers since their rate of water transfer is low compared with that of paper and fibreboard, which have a high permeability to water vapour. Nevertheless, even the use of fibreboard packages or paper bags will substantially reduce water loss compared with unprotected, loose produce. It must, however, be remembered that packaging also reduces the rate of cooling by restricting air movement around individual items.

The use of very thin plastic wrap and heat-shrink films for packaging individual fruits is a relatively underused technology that can significantly increase the storage life of many produce by greatly reducing their rates of water loss. Such films offer the benefit of storage in a water vapour-saturated atmosphere without the associated problems of condensation. Theoretically, condensation cannot occur because the film is very thin (e.g. 10–50 μm) and the produce and the film are in intimate contact and, therefore, at exactly the same temperature. In reality, there are often areas where the produce and the film are not in contact (e.g. stem scars). Condensation can occur in these positions and, in turn, fungal development can be promoted. Thus, fruit such as citrus and netted melons are often treated with approved fungicides before packaging in films. The spread of decay from any fruit that becomes infected is, however, prevented by film packaging. In the case of citrus fruit there is evidence that healing of skin injuries inflicted during harvesting and packinghouse operations is promoted by the saturated atmosphere created within the heat-shrink film. If the film is not fully sealed because of incomplete coverage, inadvertent holes and/or deliberate macro- or microperforations there is little change in the composition of the internal respiratory gas atmosphere in the fruit, and therefore little risk of anaerobiosis and associated quality defects (e.g. off-flavours, dis-colouration). Since the heat-shrink film is in direct contact with the surface of individual fruit there is little effect on the rate of heat exchange.

The ability of many packaging materials to absorb water warrants consideration. Paper derivatives, jute (hessian) bags and natural fibres can absorb a considerable amount of water before becoming visibly damp. At the time of packing there is often a significant vpd between the produce and the package. Thus, water is evaporated from the produce and absorbed by the packaging material. In the cool storage of apples and pears it has been found that a 'dry' wooden box weighing 4 kg can absorb about 500 g of water at 0°C. Ideally, packages should be equilibrated at high humidity before use, but this is impractical commercially. An alternative procedure is to waterproof hydrophilic packaging material such as fibreboard by the use of waxes or resins or to use a non-absorbent material such as polystyrene.

From a different perspective, the ability of packaging and other materials (e.g. inorganic salts in sachets permeable to water vapour but not to free water) to absorb water and/or water vapour can be used to achieve partial moisture control within packages. Such moisture sinks may be utilised to lower RH and avoid condensation within a package and, in turn, to reduce disorders such as fruit splitting or decay. Conversely, moisture sinks with sufficient absorption capacity can also work as reservoirs or moisture stores. Such moisture stores can act to return water vapour to produce that is dehydrating within a package. For example, in the case of packaged roses, dry paper packaging can be effectively used to inhibit the growth and development of grey mould. On the other hand, wrapping roses in moistened paper can effectively reduce bent neck, a disorder associated with water loss from the rose flower peduncle.

## ORNAMENTALS

Unlike fruit and nearly all vegetables, ornamentals are often provided with water for part (e.g. cut flowers and foliage) or all (e.g. pot plants) of the post-production period. In the case of most pot plants, it is important to keep the growing medium moist to avoid wilting and water stress of the plant rooted in that medium. To this end, ingredients with high water-holding capacity are usually included in the rooting medium (e.g. vermiculite, peat, hydrophilic synthetic polymers). However, during periods in confined spaces (e.g. in transit), reduction of

RH (e.g. by ventilation) is essential in order to minimise fungal and bacterial diseases.

Termination of vase life in fresh cut flowers and foliage is often associated with wilting, whereby, as for pot plants, transpiration exceeds the supply of water. Provided that there is solution in the vase and that the ambient conditions are not extremely adverse, such as for flowers placed in direct sunlight (high temperature) in an air conditioned room (low RH), wilting occurs because of increased resistance to water flow through the stem. Resistances to stem water flow can be classed as physical, physiological or biological in nature. Physical resistance includes air bubbles entrapped in xylem cells at the cut end of the stem and plugging with inorganic particles such as clay and organic particles such as live and dead bacteria and cell wall fragments. Influx of air into cut xylem endings can be avoided by recutting the stem ends under water. There have been suggestions that cold water (e.g. <5°C) can dissolve more air per unit volume than warm water, and will therefore reduce the effects of air bubbles on water flow through the stem. However, this argument applies more to removing restrictions on water flow through the microcapilliaries of the cell walls of vascular tissue in the stem than to facilitating water flow through the lumens of xylem vessels. As a result of a significant increase in stem resistance to water flow, the column of water above the critical resistance will come under increasing tension. In turn, cavitation may occur, with the consequence that resultant air emboli serve to further block stem flow. Physiological plugging implies the involvement of active metabolism by the cut stem in processes that seal off the cut stem end (wound), such as in the formation of tyloses in the lumen of xylem vessels. Biological plugging recognises that, in addition to physical plugging by both live and dead microbes, some live microbes might actively alter the structure of the water-conducting plant tissue (i.e. digest it). Factors that may determine the relative detrimental effects of different microbes include size, the formation of extracellular slime matrices, and the secretion of enzymes (e.g. pectinases) and plant toxins. Nevertheless, it is often observed that cut flower and foliage stems with prolific microbial growth on and around the cut end in the vase solution still retain turgor. In such instances it is probable that plugging is incomplete, stomata are closed or their aperture much reduced, and/or water flow occurs via alternative routes such as by cracks or lenticels in the epidermis of the stem and via symplastic and/or cell wall pathways. Cut flower and foliage species with characteristically high rates of water use (e.g. roses) appear to be relatively more susceptible to wilting.

A wide variety of chemicals are included in vase solutions in order to help maintain a positive water balance. Acidifiers such as citric acid are used with a view to specifically enhancing water flow (although the mechanism of action is unknown) and to reduce the pH below that which will support the growth of most microbes. Biocides such as chlorine and quaternary ammonium compounds are used specifically to kill microbes. Chemicals that suppress ethylene synthesis or action (e.g. silver thiosulphate, or STS) may possibly inhibit physiological plugging. Abscisic acid (ABA) has been used to induce stomatal closure and thereby increase resistance to water loss from leaves and other organs bearing stomata. ABA and other anti-transpirants may be applied as a dip or spray. Compatible inorganic (e.g. KCl) and organic (e.g. sucrose) osmotica can be provided in vase solutions to lower the osmotic potential of cells to favour water retention by cells and maintain turgor. Surfactants, which apparently decrease resistance to water flow in the xylem and which may promote dissolution of air embolisms, can also be used in vase solutions to preserve the water balance of cut flowers and foliage. It is important to determine effective concentrations and to avoid toxicity with each of the chemicals used to maintain the water balance of specific fresh cut flower and foliage lines.

It is very important to maintain a high water status during the transport of roses. The elongating peduncle joining the rose bud to the stem contains a fragile section that is at best poorly lignified. In this section, turgor is of primary importance in supporting the relatively heavy rose bud. If turgor is lost and the bud moves around during handling, the cells in the fragile part of the peduncle may be badly bruised. Typically, the damaged section blackens and cannot support the rose bud, giving rise to the disorder known as bent neck. On the other hand, it is common practice to wilt kangaroo paw inflorescences prior to packaging and transport. As a result of increased flexibility, fewer flowering cymes are snapped off the inflorescence during handling. Wilted kangaroo paw inflorescences readily rehydrate, with little or no apparent loss in vase life.

Water considerations are also something of an issue with preserved ornamental materials. Humectants, such as glycerol, are impregnated into plant material either by immersion into vats of solution or by uptake of solution via the transpiration stream. These humectants attract water vapour from the atmosphere, which helps maintain otherwise dry plant material in a supple (plasticised) condition. However, if preserved material is over-glycerined it can sweat droplets of glycerol solution at high RH. Apart from being unsightly and sticky, sweating also supports the growth of moulds.

# FURTHER READING
......

Alam, S., D. Joyce and A. Wearing (1996) Effects of equilibrium relative humidity on *in vitro* growth of *Botrytis cinerea and Alternaria alternata*, *Aust. J. Exp. Agric.* 36: 383–88.

ASHS (1978) *Relative humidity — Physical realities and horticultural implications: Proceedings of the symposium, 14 October 1977*, Salt Lake City, UT, HortScience, 13: 549–74.

Beek, van, G. (1985) Practical applications of transpiration coefficients of horticultural produce, *ASHRAE Trans.* 91(1B): 708–25.

Ben-Yehoshua, S. (1987) Transpiration, water stress and gas exchange, in J. Weichmann (ed.) *Postharvest physiology of vegetables*, Marcel Dekker, New York.

Dennis, C. (1986) Post-harvest spoilage of fruits and vegetables, in P.G. Ayres and L. Boddy (eds) *Water, fungi and plants*, Cambridge University Press, Sydney, pp. 343–57.

Fockens, F.H. and H.F. Meffert (1972) Biophysical properties of horticultural products as related to loss of moisture during cooling down, *J Sci. Food Agric.* 23: 285–98.

Gaffney, J.J., C.D. Baird and K.V. Chau (1985) Influence of air flow rate, respiration, evaporative cooling, and other factors affecting weight loss calculations for fruit and vegetables, *ASHRAE Trans.* 91(1B): 690–707.

Grierson, W. and W.F. Wardowski (1978) Relative humidity effects on the postharvest life of fruits and vegetables, *HortScience*, 13: 570–74.

Halevy, A.H. and S. Mayak (1979) Senescence and postharvest physiology of cut flowers, Part 1, *Hort. Rev.* 1: 204–36.

Halevy, A.H. and S. Mayak (1981) Senescence and postharvest physiology of cut flowers, Part 2, *Hort. Rev.* 3: 59–143.

Lazan, H., Z.M. Ali, A.A. Mohd and F. Nahar (1987) Water stress and quality decline during storage of tropical leafy vegetables, *J. Food Sci.* 52: 1286–88, 1292.

Milburn, J.A. (1979) *Water flow in plants*, Longman, London.

Nobel, P. (1974) *Introduction to biophysical plant physiology*, W.H. Freeman, San Francisco, CA.

Sastry, S.K. (1985) Moisture losses from perishable commodities: Recent research and developments, *Int. J. Refrig.* 8: 343–46.

Sharp, A.K. (1986) Humidity: Measurement and control during the storage and transport of fruit and vegetables, *CSIRO Food Res. Quart.* 46: 79–85.

Thompson, A.K. (1996) *Postharvest technology of fruit and vegetables*, Blackwell Science, Oxford.

Tyree, M.T. and P.G. Jarvis (1982) Water in tissues and cells, in O.L. Lange, P.S. Nobel, C.B. Osmond and H. Ziegler, *Physiological plant ecology*, vol. II, *Water relations and carbon assimilation*, Springer-Verlag, Berlin.

Weichmann, J. (1987) *Postharvest physiology of vegetables*, Marcel Dekker, New York.

# 6
# STORAGE ATMOSPHERE

The composition of gases in the storage atmosphere can affect the storage life of horticultural produce. Changes in the concentrations of the respiratory gases — oxygen and carbon dioxide — may extend storage life. This is generally used as an adjunct to low temperature storage, but modification of the storage atmosphere can usefully substitute for refrigeration for some commodities. Many volatile compounds released by produce and from other sources may accumulate in the storage atmosphere. Ethylene is the most important of these compounds and its accumulation above certain critical levels reduces storage life, so methods for its control become important. A role in quality maintenance for other organic volatiles such as acetaldehyde and ethanol released by produce is currently under investigation, particularly in relation to antimicrobial properties for certain compounds.

The terms controlled atmosphere (CA) storage, modified atmosphere (MA) storage and gas storage are frequently used. These terms imply the addition or removal of gases resulting in an atmospheric composition different from that of normal air. Thus the levels of carbon dioxide, oxygen, nitrogen and ethylene in the atmosphere may be manipulated. Controlled atmosphere storage generally refers to decreased oxygen and increased carbon dioxide concentrations, and implies precise control of these gases. The term modified atmosphere storage is used when the composition of the storage atmosphere is not closely controlled, such as in plastic film packages where the change in the composition of the atmosphere occurs intentionally or unintentionally. A more recent term is modified atmosphere packaging (MAP), which relates to packages and film box liners with specific properties that offer a measure of control over the composition of the atmosphere around produce. The original term, gas storage, is now considered inappropriate and should not be used.

## CARBON DIOXIDE AND OXYGEN

The general equation for produce respiration is:

glucose + oxygen → carbon dioxide + water + energy

It suggests that respiration could be slowed by limiting the oxygen or by raising the carbon dioxide concentration in the storage atmosphere. The principle appears to have been applied in ancient times, even if unwittingly. The earliest use of modified atmosphere storage may possibly be attributed to the Chinese. Ancient writings report that litchi was transported from southern China to northern China in sealed clay pots to which fresh leaves and grass were added. It may be surmised that during the 2 week journey, respiration of the fruit, leaves and grass generated an atmosphere in the pots that was high in carbon dioxide and low in oxygen, which retarded ripening of litchi. Other examples of primitive modified atmosphere storage include the burying of apples in the ground and the carriage of fruit in the unventilated holds of ships. The first reported scientific observations of the effects of the atmosphere on fruit ripening were made in 1819–20 by Jacques Berard in France, but it was not until the work of Kidd and West at the Low Temperature Research Station at Cambridge, UK, in the 1920s and 1930s that a sound basis for the controlled atmosphere storage of produce was established.

By showing that high carbon dioxide and low oxygen concentrations lowered the respiration rate of seeds and delayed their germination, Kidd and West extended earlier findings that the composition of the atmosphere affected the metabolic rate of plant tissues. It was soon apparent to them that there should be a similar effect in fruit. They subsequently concentrated their work on apples, which culminated in the publication in 1927 of the classic bulletin entitled *Gas Storage of Fruit*. An important stimulus to the work of Kidd and West was the fact that the commercial cultivars grown in the UK were subject to low temperature disorders when stored at less than 3°C and that storage life was too short at temperatures above 3°C.

During the past 50 years, the effects of controlled atmosphere and modified atmosphere have been extensively tested on a wide range of fruit and vegetables, but the responses have varied considerably. Despite extensive research, major commercial application of controlled atmosphere has been confined to some apple and pear cultivars, but modified atmospheres have been applied successfully during transport to a range of produce. For example, high carbon dioxide levels have been used primarily as a fungistat during the transport of strawberries, and improved outturns of lettuce have been achieved by flushing rail trucks or containers with nitrogen and up to 8 per cent carbon monoxide.

As with fruit and vegetables in general, controlled atmosphere storage of ornamentals has not been widely implemented. Nonetheless, controlled atmosphere recommendations have been established for a number of important cut flower crops (Table 6.1). In addition to extended longevity at low temperatures, storage under controlled atmosphere conditions, particularly under high carbon dioxide concentrations, can limit the development of pathogens. However, the lack of commercial usage arises from considerably inconsistent responses between different seasons and different varieties within a given species. In addition, regular production schedules of major species and limited quantities of minor species create little need to store many cut flowers for significant periods.

### TABLE 6.1  OPTIMAL CONTROLLED ATMOSPHERE CONDITIONS FOR CERTAIN FLOWERS

| Species | %$CO_2$ | %$O_2$ | Temp (°C) | Storage period |
|---|---|---|---|---|
| Freesia | 10 | 21 | 1 | 3 weeks |
| Carnation | 5 | 2 | 0 | 4 weeks |
| Lily | 10–20 | 21 | 1 | 3 weeks |
| Mimosa | 0 | 8 | 7 | 10 days |
| Rose | 5–10 | 1–3 | 0 | 3–4 weeks |

Source  Adapted from D.M. Goszczynska and R.M. Rudnicki (1988) Storage of cut flowers, *Hort. Rev.* 10: 35–62.

Factors that have influenced the adoption of controlled or modified atmospheres for different commodities include:

◆ Inherent storage life in air. If the produce can be stored in a satisfactory condition in air for the total marketing period desired, then there is no need to resort to atmosphere modification to prolong storage life.

◆ Existence and magnitude of a favourable response to modified atmospheres. There must be a distinct beneficial effect. Not all produce responds favourably to atmosphere regulation and some produce is little affected.

◆ Substantial atmosphere tolerance. The beneficial effects of atmosphere modification, especially in modified atmosphere storage, need to be sustained over a relatively wide gas concentration range. A small tolerance range will result in variable quality outturns in commercial usage due to insufficient or excessive gas concentrations.

◆ Seasonal availability. Use of atmospheres can be advantageous where produce is harvested over a relatively short period in the year. Maximum storage life of such produce is often desirable to extend the marketing period over a greater part of the year.

♦ Value of the commodity in relation to the cost of atmosphere modification. There needs to be a distinct financial gain from the use of atmosphere control.

♦ Availability of substitute commodities. While produce may be stored satisfactorily in modified atmospheres, it may be more economical to import produce from another region or country that has a different harvest period.

The use of modified atmospheres on portable packages should have considerable commercial appeal, but the lack of widespread application is largely related to problems in ensuring that the desired atmosphere is maintained throughout the postharvest chain under a diverse range of handling operations and external environmental conditions. A variable modified atmosphere will often result in variable produce quality in the market. The availability of a range of new food-grade polymeric films with different permeabilities to the atmospheric gases in recent years has revived interest in packaging produce in sealed bags. The use of such films for refrigerated produce offers a cheaper alternative to using large containers equipped to provide modified or controlled atmosphere conditions. The development and properties of these plastic films will be covered in greater detail in Chapter 12.

## Metabolic effects

Increases in carbon dioxide and decreases in oxygen concentrations exert largely independent effects on respiration and other metabolic reactions. Generally, the oxygen concentration must be reduced to less than 10 per cent (by volume) before any retardation of respiration is achieved. For apples stored at 5°C, the oxygen level must be reduced to about 2.5 per cent to achieve a 50 per cent reduction in the respiration rate. Care must also be taken to ensure that sufficient oxygen is retained in the atmosphere so that anaerobic respiration, with its associated development of off-flavours, is not initiated.

The reduction in the concentration of oxygen necessary to retard respiration depends on the storage temperature. As the temperature is lowered, the required concentration of oxygen is also reduced. The critical level of oxygen at which anaerobic respiration occurs is determined mainly by the rate of respiration, and is therefore greater at higher temperatures. Tolerance to low oxygen levels varies considerably among different commodities. The critical level of oxygen may vary with the time of exposure, with lower levels being tolerated for shorter periods. It may also be affected by the level of carbon dioxide, since lower levels of oxygen often seem to be better tolerated when carbon dioxide is absent or at a low level.

The addition of only a few per cent of carbon dioxide to the storage atmosphere can have a marked effect on respiration. However, if carbon dioxide levels are too high, effects similar to those caused by anaerobiosis (lack of oxygen) can be initiated. Responses to increased carbon dioxide levels vary even more widely than responses to reduced oxygen: cherries and strawberries will withstand, and even benefit from, exposure to 30 per cent carbon dioxide for short periods. In contrast, some apple cultivars are injured by 2 per cent carbon dioxide in storage, and many vegetables appear to respond best to low oxygen when carbon dioxide is kept low or is absent (Figure 6.1).

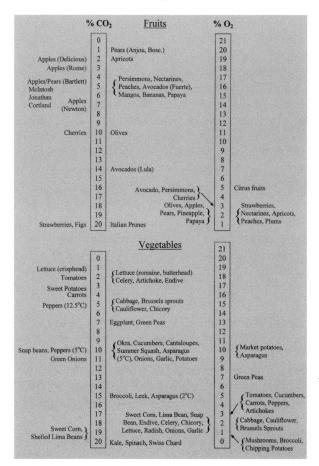

**Figure 6.1**

Relative tolerance of fruit and vegetables to elevated carbon dioxide and reduced oxygen concentrations at recommended storage temperatures. Normal atmospheric air comprises 0.035 per cent carbon dioxide, 21 per cent oxygen and about 79 per cent nitrogen. SOURCE A.A. Kader and L.L. Morris (1977) Relative tolerance of fruits and vegetables to elevated $CO_2$ and reduced $O_2$ levels, in D.H. Dewey (ed.) *Controlled atmospheres for storage and transport of horticultural crops,* Michigan State University, East Lansing, MI, 260–65.

Many of the beneficial results of modified atmosphere storage cannot simply be attributed to a reduction in respiration. For example, under ideal experimental conditions a 12-fold increase in the storage life of green bananas can be achieved by ventilating the fruits with an atmosphere comprising 5 per cent carbon dioxide, 3 per cent oxygen and 92 per cent nitrogen in the absence of ethylene, but respiration measured in terms of oxygen uptake is reduced to only one-quarter of the rate in air. The greatly increased storage life is attributed to a reduction in the rate of natural ethylene production by the bananas and also to a reduced sensitivity of the fruits to ethylene.

In green vegetables, improved retention of green colour in low oxygen atmospheres is due mainly to a lowering of the rate of chlorophyll destruction. An interesting and contrasting effect has been noted in potato. Greening due to exposure to light can be prevented for several days by maintaining tubers in an atmosphere containing about 15 per cent carbon dioxide.

Modified atmospheres, particularly those containing high carbon dioxide, inhibit the breakdown of pectic substances so that a firmer texture is retained for a longer period. Retention of flavour may also be improved. The responses of various commodities to modified atmospheres can, however, be conflicting. For example, increased carbon dioxide aids in the retention of organic acids in tomato but accelerates the loss of acids in asparagus. Maturity at harvest is more critical for modified atmosphere storage than for ordinary air storage. Because of the widely varying responses of different commodities, and among cultivars, to alterations in oxygen and carbon dioxide concentrations, ideal combinations need to be determined experimentally for each commodity.

### Effect on microbial growth

The activity of several decay organisms can be reduced by atmospheres containing 10 per cent carbon dioxide or more, provided that the commodity is not injured by such high carbon dioxide levels. Since strawberries can tolerate high carbon dioxide levels, transport of strawberries under modified atmosphere conditions has been found to significantly reduce rotting and give a valuable extension in marketing life and quality (Table 6.2).

Many commodities cannot tolerate high carbon dioxide levels so that, in practice, atmosphere control cannot always be relied on to reduce rotting. Nevertheless, atmosphere control (by increased carbon dioxide and reduced oxygen) may reduce rotting of produce by retarding ripening and senescence, since the natural resistance of the produce host to pathogens decreases as it ripens or ages. In contrast, some fruits, such

### TABLE 6.2 DECAY OF STRAWBERRIES AS INFLUENCED BY CARBON DIOXIDE CONCENTRATION (% OF STRAWBERRIES DECAYED)

| STORAGE CONDITION | 0% $CO_2$ | 10% $CO_2$ | 20% $CO_2$ | 30% $CO_2$ |
|---|---|---|---|---|
| 3 days at 5°C in storage atmosphere | 11.4 | 4.5 | 1.7 | 1.3 |
| Plus 1 day at 15°C in air | 35.4 | 8.5 | 4.7 | 4.0 |
| Plus 2 days at 15°C in air | 64.4 | 26.2 | 10.8 | 8.3 |

SOURCE Adapted from C.M. Harris and J.M. Harvey (1973) Quality and decay of California strawberries stored in carbon dioxide-enriched atmospheres, *Plant Dis. Reporter*, 57: 44–6.

as banana and mango, respond well to atmosphere control but eventually lose their resistance to the latent anthracnose disease, which then becomes the factor limiting storage life. Controlled atmosphere or modified atmosphere storage does not necessarily retard ageing and the loss of resistance to decay organisms at the same rates.

Carbon monoxide is an effective fungistat but care must be exercised in its use. As well as being explosive, carbon monoxide is a potent mammalian poison binding to blood haemoglobin. It also has the ability to mimic the effects of ethylene when used in the absence of carbon dioxide.

## Methods for modifying carbon dioxide and oxygen concentrations

The first commercial applications of controlled atmosphere storage relied on the produce to generate the atmosphere so that carbon dioxide concentrations approximately equalled the reduction in oxygen. The composition of the storage atmosphere was generally maintained in the range 5–10 per cent carbon dioxide and 16–11 per cent oxygen. The build-up of carbon dioxide was mainly responsible for the increased storage life. The store was ventilated regularly to maintain the required carbon dioxide level. Further research showed that low oxygen levels were of greater benefit and that some important cultivars of apple and pear were sensitive to carbon dioxide levels above 3 per cent. To maintain low oxygen levels in the store atmosphere it was necessary to make existing cool stores more gas-tight and to recirculate part of the store atmosphere through a scrubber to remove excess carbon dioxide. Gas-tight cool stores were then developed for controlled atmosphere storage and, along with external generators, increased the viability of commercial controlled atmosphere storage. These generators burn fuels such as propane or petroleum gas and rapidly reduce the oxygen in the store to a required low level that is then maintained. Excess carbon dioxide is removed by scrubbing equipment. Gas generators also allow controlled atmosphere storage to be accomplished, albeit expensively, in a store that is not completely gas-tight.

As a result of these developments in cool store design and operation, the more effective low oxygen atmospheres containing 2–5 per cent carbon dioxide and 2–3 per cent oxygen can readily be maintained. The oxygen concentration must be carefully controlled to prevent anaerobic respiration or fermentation in this type of atmosphere. Methods of constructing and maintaining controlled atmosphere stores are discussed in Chapter 7.

### Atmosphere control by the addition of nitrogen and carbon dioxide

In recent years there has been renewed interest in atmosphere control in the long distance transport of perishable produce in containers. One of the factors responsible for this interest is the availability in some countries, notably the USA, of cheap liquid nitrogen. Atmosphere control in large containers has involved either the continuous introduction of carbon dioxide or nitrogen gas during the journey, or charging the container with the appropriate atmosphere before the journey, with no further introduction of gas. Carbon dioxide or nitrogen from pressurised cylinders are used, depending on whether the requirement is for high carbon dioxide or low oxygen levels, or both.

The advent of hollow fibre systems for separating oxygen and nitrogen in air by differential diffusion across a membrane has provided a new means of producing low oxygen atmospheres for continuous ventilation of produce in large containers or in fixed storage rooms. An advantage of the technique is that the starting raw material, normal air, is freely available and there are no hazardous gaseous by-products generated.

The use of liquid nitrogen as a refrigerant in the transport of perishables had stimulated interest in the response of fruit and vegetables to very high nitrogen levels and consequent concentrations of oxygen of 1 per cent or less, both under refrigeration and at higher temperatures. It is now known that many fruit and vegetables can withstand such atmospheres for a short period without harm and show a long-term reduction in respiration when returned to storage in air. The beneficial effects are noted particularly for non–climacteric fruit and vegetables. It also means that the use of liquid nitrogen refrigeration in road and rail transport vehicles is unlikely to be damaging to most perishables because the likelihood of maintaining oxygen-free atmospheres is remote and even 1 per cent oxygen is enough for most commodities to remain viable for several days.

## Storage in plastic films

The use of plastic films to achieve modified atmosphere is increasing, not only in packaging but also in controlled atmosphere store construction (Chapter 7). Polyethylene box liners, either sealed or unsealed, have been used for several years in the storage of pears and apples but only to a limited extent with other produce. Unsealed or perforated bags are commonly used to minimise weight loss and reduce abrasion damage. A major problem with sealed bags is that the atmosphere inside depends on the temperature, because the permeability of the film to gases is virtually independent of temperatures at which produce is normally handled whereas respiration is temperature-dependent. Thus sealed bags are risky when the temperature varies more than a few degrees, unless the produce has a low rate of respiration or is tolerant to atmospheres that vary widely in carbon dioxide and oxygen concentrations (like the banana), or both. The film commonly used is 40 μm (0.0015 inch) low density polyethylene. To avoid brown spot and other carbon dioxide injuries, sachets of fresh hydrated lime can be included in the bag to reduce the carbon dioxide concentration. At the rate of 100–200 grams per 10 kilograms of fruit, this has proved useful with apples and pears, particularly with carbon dioxide-sensitive cultivars in cool storage.

The attainment of modified atmosphere in polyethylene bags filled with produce can be accelerated by evacuating the bags to between 50 and 85 kPa (380–635 mm mercury) and then sealing them. Since the polyethylene film is permeable to nitrogen, oxygen and carbon dioxide, the pressure inside returns to atmospheric pressure, but the initial rapid reduction of oxygen concentration is often useful. Eventually the composition of the atmosphere approaches that in bags not subjected to initial evacuation. Fruit must be removed from the bags to achieve normal ripening. If held for longer periods under modified atmosphere, the fruits may not ripen satisfactorily after removal.

The interest in plastic films to generate modified atmospheres has accelerated in recent years through the availability of films with more flexible gas permeabilities. The newer films remove many of the risks of modified atmosphere storage outlined above. The types of film will be discussed in Chapter 12.

## Hypobaric storage

Hypobaric storage is a form of controlled atmosphere storage in which the produce is stored in a partial vacuum. The vacuum chamber is vented continuously with water-saturated air to maintain oxygen levels and to minimise water loss. Ripening of fruit is retarded by hypobaric storage due to the reduction in the partial pressure of oxygen and for some fruits also due to the reduction in ethylene levels. A reduction in pressure of air to 10 kPa (0.1 atmosphere) is equivalent to reducing the oxygen concentration to about 2 per cent at normal atmospheric pressure. Hypobaric stores are expensive to construct because of the low internal pressures required, and this high cost of application appears to limit hypobaric storage to high value produce. The technique has now found some application in the storage of fresh meats.

# ETHYLENE
.......

## *Effects on fruit and vegetables*

The commencement of natural ripening in climacteric fruits is accompanied by an increase in ethylene production (Chapter 3). Treatment of pre-climacteric fruits with exogenous ethylene advances the onset of ripening. This response is used widely in commercial practice to achieve controlled ripening of fruit such as banana, which is picked and transported in a mature but unripe state and ripened just before marketing (Chapter 11). The action of ethylene must, however, be avoided for such fruit during storage and transport to prevent premature ripening. In contrast, the effect of ethylene on non-climacteric fruit and vegetables offers no apparent commercial benefit but will reduce postharvest quality by promoting senescence, as evidenced by loss of green colour, change in texture and flavour, and promotion of low temperature injuries and microbial decay.

While the levels of ethylene that trigger ripening have been well researched for most climacteric fruits, the threshold concentration that enhances senescence in non-climacteric fruit and vegetables is less well documented. A concentration of 0.1 µL/L is often cited as the threshold level, but recent studies in Australia indicate that the threshold level of ethylene is less than 0.005 µL/L. For practical purposes, this means there is no safe level of ethylene, and hence any reduction in ethylene concentration will bring some extension in postharvest life.

Ethylene in a storage or transport container may come from produce or from outside sources. Often during marketing, several commodity types are stored together, and under these conditions ethylene given off by one commodity can adversely affect another. Coal gas, petroleum gas and exhaust gases from internal combustion engines contain ethylene, and contamination of stored produce by these gases may introduce sufficient ethylene to initiate ripening in fruit and promote deterioration in non-climacteric produce and ornamentals. The level of deleterious action will depend on the concentration of ethylene that accumulates and the duration of exposure.

In addition to delaying ripening or senescence through reducing ethylene concentrations around produce, the sensitivity of produce to ethylene may be lessened by storage at low temperature, and by either raising the level of carbon dioxide or decreasing the level of oxygen. Under these conditions the amount of ethylene required to induce ripening is increased. A similar effect has been demonstrated for some non-climacteric produce. For example the ethylene-induced breakdown of chlorophyll in broccoli is less sensitive to ethylene at low temperatures.

## Effects on ornamentals

Many ornamental crops are sensitive to ethylene. Responses of ornamentals to ethylene can be classed into growth, abscission and senescence responses. An example of an ethylene-induced growth response is epinastic curvature of poinsettia leaves and bracts. Abscission is a far more widespread response across the broad range of ornamental species. Many types of organs may abscise, including stem segments, leaves, fruit, whole inflorescences, buds and flowers, and petals. For example, fruit, leaves and stem segments of Christmas mistletoe sprigs (*Phoradendron tomentosum*) all separate upon exposure to ethylene. Ethylene-induced accelerated senescence is characterised by premature discolouration and wilting of flowers, such as carnations and cymbidium orchids.

Individual species vary widely in their relative sensitivity to ethylene and, in general, cut flowers and flowering pot plants tend to be more ethylene-sensitive than foliage lines. Among cut flowers, carnation and delphinium are considered to be very sensitive to ethylene, while gerbera and tulip are considered relatively insensitive. Among flowering pot plants, hibiscus is classed as highly sensitive and chrysanthemum is of low sensitivity. Finally, among foliage plants, schefflera is highly sensitive and nephrolopis is insensitive. However, it must be noted that ethylene sensitivity can vary markedly among genotypes (e.g. species) within a genus.

Similarly, ethylene production can vary widely between genotypes. For example, some carnations produce a marked ethylene climacteric peak during senescence, whereas others do not produce significant amounts of ethylene. As has been observed for genotype, factors that influence phenotype — including preharvest temperature, RH, light and nutrition regimes — can also affect the relative sensitivity of ornamentals to ethylene.

# METHODS FOR REDUCING ETHYLENE CONCENTRATIONS

......

## *Avoidance of ethylene accumulation*

Reduction of ethylene levels in storage rooms can be achieved by good housekeeping; that is, storing ripe and unripe produce in separate rooms, regularly removing all rotted or damaged produce, and by ensuring that natural gas pipes and cylinders, and exhaust gases from internal combustion engines, are kept well away from storage rooms.

A simple physical method to minimise ethylene accumulation is to ensure good ventilation of the storage chamber with air from outside the storage complex. The ethylene concentration in the atmosphere is normally less than 0.005 µL/L unless there is contamination from nearby industrial sources or heavy automobile traffic. Ventilation with external air could be applicable where no large temperature differential exists between the external air and the storage chamber provided it is at an appropriate relative humidity. If there is a large temperature difference it may be necessary to cool the air before admitting it to the chamber.

## *Oxidation with potassium permanganate*

Ethylene in the atmosphere can be oxidised to carbon dioxide and water using a range of chemical agents. Potassium permanganate ($KMnO_4$) is quite effective in reducing ethylene levels. Since it is non-volatile, potassium permanganate can be physically separate from produce, thus eliminating the risk of chemical injury. To ensure efficient destruction of ethylene, a large surface area of potassium permanganate is achieved by coating an inert inorganic porous support, such as alumina or expanded mica, with a saturated solution of potassium permanganate. Potassium permanganate used in this manner has been found to retard the ripening of many fruits. Table 6.3 demonstrates the benefit to be obtained with banana when used in conjunction with modified atmosphere storage in polyethylene bags. The high carbon dioxide and low oxygen atmosphere generated within the sealed bags decreases the response by the fruits to ethylene and hence retards ripening. The addition of potassium permanganate further retards ripening by maintaining ethylene at a low level for a long period. The technique has also been successfully demonstrated to delay the ripening of whole bunches of bananas during growth on the plant. High humidity in storage containers is a limitation to the longevity of potassium permanganate absorbents as it also reacts with water. While the commercial use of potassium permanganate alone or in conjunction with modified atmosphere is difficult to document, it must be widespread in many countries to sustain the various commercial manufacturers of potassium permanganate sachets that remain in business.

### TABLE 6.3 SHELF LIFE OF BANANAS HELD AT 20°C IN MODIFIED ATMOSPHERE WITH POTASSIUM PERMANGANATE

| TREATMENT | SHELF LIFE (DAYS) |
|---|---|
| Air | up to 7 |
| Sealed polyethylene bags | 14 |
| Sealed bags + potassium permanganate | 21 |

SOURCE Derived from K.J. Scott, W.B. McGlasson and E.A. Roberts (1970) Potassium permanganate as an ethylene absorbent in polyethylene bags to delay ripening of bananas during storage, *Aust. J. Exp. Agric. Anim. Husb.* 10: 237–40.

## Oxidation with ozone

Ozone ($O_3$) is a suitable oxidising agent for destroying ethylene because it is generated readily from atmospheric oxygen by an electric discharge or ultraviolet radiation, and since it is gaseous, it readily mixes with ethylene. The actual oxidant species is probably a combination of ozone and atomic oxygen, a highly reactive free radical formed from ozone. Some precautions must be taken with ozone: it is a reactive substance and will corrode metal pipes and fittings in refrigeration equipment and react with paper products used to package produce. In addition, ozone readily injures produce and can be toxic to humans at relatively low concentrations. The widespread use of ozone has been hampered by difficulties in controlling its concentration. These problems with ozone can be overcome by its use in a recycling system, as depicted in Figure 6.2. Here ozone is generated in a container with ultraviolet radiation. Air contaminated with ethylene is passed through the chamber where the ethylene is oxidised, and excess ozone is removed by reduction on a substrate such as steel wool. Some small scale commercial ultraviolet scrubbers have been produced but are not in widespread use.

## Other oxidants

Activated charcoal that has been brominated will effectively oxidise ethylene but carries a potential health hazard as it generates bromine gas when in contact with excess water. Recent laboratory studies have identified a range of chemicals, such as tetrazine, that react more specifically with ethylene. Their greater specificity for ethylene makes them relatively more efficient in ethylene scavenging than general oxidising agents such as potassium permanganate. Commercial applications are focusing on their inclusion into plastic packaging films. A problem to be overcome is the instability of tetrazines in the presence of moisture.

## Other chemicals

A range of chemicals that are toxic to humans can be used as anti-ethylene treatments with ornamentals because they are not eaten. Most notably, cut stems of ornamental material can be pulsed with silver ions supplied as STS, or whole plants or plant parts can be dipped or sprayed with STS. Silver ions block the ethylene binding site, thereby preventing its action. A relatively new ethylene binding site blocker, 1-methylcyclopropene (1-MCP), is likely to replace STS as an anti-ethylene treatment, and is particularly exciting as it can be applied as a gas. At this stage 1-MCP is considered safe in that it has low or no mammalian toxicity and low or no adverse environmental impact. Its application to fruit and vegetables is in relatively early stages of evaluation.

Treatment with ethylene synthesis inhibitors is an alternative to treatment with ethylene binding site blockers. Two such compounds are aminoethoxyvinylglycine (AVG) and aminooxyacetic acid (AOA). Both compounds are used to treat ornamentals, such as carnations, against ethylene. However, the relative drawback is that they only confer protection against endogenously produced, and not exogenous, ethylene. Another approach to reducing or eliminating the effects of ethylene on ornamentals is by genetic engineering. Carnation genotypes that formerly produced ethylene during senescence (i.e. climacteric

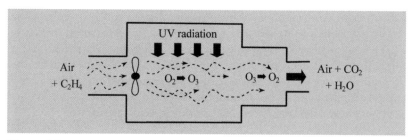

**Figure 6.2**
System for removal of ethylene with atomic oxygen generated by ultraviolet radiation.
SOURCE Derived from K.J. Scott and R.B.H. Wills (1973) Atmospheric pollutants destroyed in an ultraviolet scrubber, *Lab. Practice*, 22: 103–106.

characteristic) have been engineered using antisense gene technology *not* to produce ethylene. The genetically engineered carnations, showing non-climacteric senescence characteristics, have markedly extended vase life. Similarly, carnation genotypes with blocked ethylene binding and signal transduction and translation will be engineered. These genetic modifications emulate the effects of the chemical ethylene synthesis (e.g. AVG) and binding (e.g 1-MCP) inhibitors, respectively.

# OTHER GASES

Carbon monoxide (CO) is not released by fresh produce, but it may be introduced to storage atmospheres by equipment powered by internal combustion engines. Carbon monoxide may reach levels that are toxic to persons working in the storage chambers, and with some produce may give effects that mimic those induced by ethylene. However, there are examples of beneficial responses to added carbon monoxide, such as the control of butt discolouration and retardation of the growth of *Botrytis* rots in lettuce. The practice of adding about 5 per cent carbon monoxide to containers or pallet loads of various perishable fruits held in controlled atmospheres is now considered advantageous and is in limited use for export shipments or long distance land transport of specific produce.

# FURTHER READING
.......

Brody, A.L. (ed.) (1989) *Controlled/modified atmosphere/vacuum packaging of food*, F. & N. Press, CT.

Calderon, M. and R. Barkai-Golan (eds) (1990) *Food preservation by modified atmospheres*, CRC Press, Boca Raton, FL.

Goszczynska, D.M. and R.M. Rudnicki (1988) Storage of cut flowers, *Hortic. Rev.* 10: 35–62.

Irving, A.R. (1984) Transport of horticultural produce under modified atmospheres, *CSIRO Food Res. Quart.* 44: 25–33.

Kader, A.A. (1986) *Modified atmospheres: An indexed reference list with emphasis on horticultural commodities*, University of California, Davis, CA. Supplement no. 4, Postharvest Horticulture Series 3.

Kader, A.A., D. Zagory and E.L. Kerbel, (1989) Modified atmosphere packaging of fruits and vegetables, *Crit. Rev. Food Sci.* Nutrition 28(1): 1–30.

Nowak, J. and R.M. Rudnicki, (1990) *Postharvest handling and storage of cut flowers, florist greens and potted plants*, Timber Press, Portland, OR.

Reid, M.S. (1992) Postharvest handling systems: Ornamental crops, in A.A. Kader (ed.), *Postharvest technology of horticultural crops*, University of California, Oakland CA, University of California Publication no. 3311, pp. 201–209.

Rooney, M.L. (ed.) (1995) *Active food packaging*, Blackie Academic, London.

Sherman, M. (1985) Control of ethylene in the postharvest environment, *HortScience* 20: 57–60.

Thompson, A.K. (1995) *Postharvest technology*, Blackwell Science, Oxford.

Whiteman, J. (1987) *Postharvest physiology of vegetables*, Marcel Dekker, New York.

# 7

# TECHNOLOGY OF STORAGE

It has long been known that the preservation of fresh fruit, vegetables and ornamental plants is governed by three major parameters:

1.  they keep better when cold;
2.  they are damaged by freezing; and
3.  they shrivel or wilt in dry air.

On the basis of this knowledge, control of the temperature and RH of the air around produce has been practised with increasing sophistication and success. A fourth parameter, the composition of the atmosphere, has been recognised more recently so that controlled or modified atmosphere storage is a more modern innovation.

## METHODS OF STORAGE
.......

### In-ground storage

Pit storage, or clamp storage, is a simple low technology on-farm technique that is still beneficially practised in some countries. Hard vegetables, such as potato, turnip and late season cabbage, are piled into pits dug into a hillside or in some other well drained position. The pits are lined with hay or straw; the produce is then covered with straw followed by 10–20 cm of sods and earth to protect it against freezing or from excess heat and to deflect rain. It is advantageous to include piped ventilation to the outside to reduce respiratory self-heating. Clamp storage has been shown to be suitable for storage of cassava for up to 2 months in the tropics (Figure 7.1). In Europe, perishable produce was initially stored in cellars and caves that, being below ground, were cooler than above-ground buildings in warm weather, and warmer in winter. These methods are still commonly practised in the People's Republic of China.

Cellars are more sophisticated forms of below-ground storage; they may be part of above-ground buildings or underground rooms, often in hillsides, where access is easier. Again, good drainage and protection from rain are

**Figure 7.1**
Cassava storage clamp. (Courtesy of R.H. Booth, Food and Agriculture Organization of the United Nations, Rome.)

essential. The performance of cellars is improved by providing controlled ventilation openings to enable cold air to enter and warm air to leave by convectional circulation when cooling is required. Although temperatures generally are not optimal, a good cellar will provide satisfactory storage for hard vegetables and long-keeping fruits such as apples.

### Air-cooled stores

These are simple insulated structures above ground, or partly underground, which are cooled by the circulation of colder outside air. When the temperature of the produce is above the desired level, and if the temperature of the outside air is lower (generally at night), air is circulated throughout the store by convectional or mechanical means through bottom inlet vents and top outlets fitted with dampers. Fans may be installed and are controlled manually, or automatically with differential thermostats. The air may be humidified, a process that can also be automated. Air-cooled stores are inexpensive to construct and operate and are still widely used for the storage of potato and sweet potato, both of which need relatively high storage temperatures to avoid accumulation of sugar and chilling injury, respectively.

Potatoes are commonly stored in bulk piles in stores with air delivery ducts under the floor, or at floor level, and with suitably spaced air outlets. A system of ventilating bulk piles or bins of onions with air provides an economical way to ensure that the outer scale leaves remain dry and free of decay. Garlic corms are similarly ventilated with air to prevent mould growth.

## Ice refrigeration

An advance on air-cooled storage was the use of natural ice as a refrigerant. The lower temperatures enabled longer storage of meat and other perishable foods, including horticultural produce. In North America and Northern and Central Europe, ice was harvested in the winter from frozen lakes and ponds and stored in insulated 'ice houses'. The melting of 1 kg of ice absorbs 325 kJ, but the considerable bulk of ice needed and the disposal of the melted water are disadvantages. The introduction of the small 'ice box' or 'ice chest' was a great advance in the domestic and small-scale commercial preservation of perishable foodstuffs. Ice produced by mechanical refrigeration has several modern commercial applications as an adjunct to refrigeration (Chapter 4).

## Mechanical refrigeration

The father of modern refrigeration was undoubtedly the Australian James Harrison. By 1851 he had designed and built the first ice-making plant in the world incorporating a small refrigeration compressor with its auxiliary equipment and ice tank at Geelong in Victoria. In 1854 Harrison was granted British Patent No. 717 for 'the production of cold by the evaporation of volatile liquids in vacuo', an invention probably equal in importance to that of the steam engine. The general principles of Harrison's design remain virtually unchanged in modern refrigeration plants. The system developed rapidly, and mechanically refrigerated cold stores, insulated with natural materials such as sawdust or cork, were operating within a few years. A shipment of frozen beef from Australia to England in 1879 was the first successful long distance shipment of perishable food by sea; soon after, the first mechanically refrigerated cool stores for apples and pears were in operation.

A refrigeration plant consists of four basic components: the compressor in which the refrigerant gas (either ammonia or halogenated hydrocarbons) is compressed (and unavoidably heated); the condenser, either air-cooled or water-cooled, in which the hot gas is cooled and condensed to a liquid; the expansion valve; and the evaporator coils, in which the liquid is permitted to boil and so remove heat from its surroundings (Figure 7.2). Fans are usually necessary to circulate the storage air over the cooling coils of the evaporator and through the stacks of produce in the store. The main agent for the transfer of heat from the interior of the store to the cooling coils is air movement, although radiation and convection may play a small part. In addition to these four basic components of a refrigeration plant and fans or correctly placed air ducts, various items of ancillary control equipment such as a liquid receiver and some means of defrosting the coils are required.

The removal of heat from the circulating air in a cool room by the evaporator coils is referred to as the 'direct expansion system'. Inevitably, the surface temperature of the evaporator coils must be lower than that of the produce to ensure that heat from all sources in a cool room is removed and the produce remains at a constant temperature. This temperature gradient is accompanied by a vpd between the produce and

**Figure 7.2**

Basic component parts of a mechanical refrigeration plant. The part of the cycle from the compressor to the thermal expansion valve operates under high pressure to enable condensation of the hot gaseous refrigerant. The evaporator coils operate at low pressure to enable the refrigerant to boil. The condenser may be either air or water cooled.

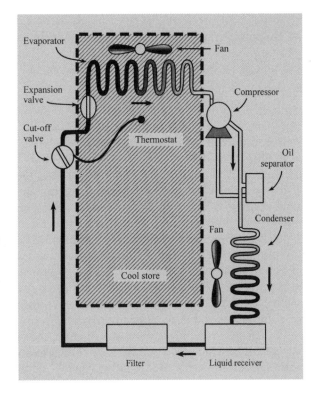

the evaporator coils. This enhances water loss from the produce. For produce with high transpiration rates — such as leafy and root vegetables, mushrooms and cut flowers — a preferred system is to cool and humidify the room air by passing it through a shower of cold water that has been cooled by mechanical refrigeration. This indirect expansion system provides air at 1–2°C and RH of more than 98%. A variation of this system allows accumulation of ice on the evaporator coils (ice bank) to augment cooling capacity at times of high demand and to take advantage of lower off-peak electricity charges if this concession is available.

# DESIGN AND CONSTRUCTION OF COOL STORES

A cool store is a large, thermally insulated box, with doors for entry and some means of cooling the interior. Cool stores for horticultural produce have special requirements in comparison with other refrigerated stores. These include a high cooling capacity, close control of temperature, and RH greater than 90%. A common minimum design criterion is to provide capacity to cool a daily intake of 10 per cent of the capacity of the store at an initial rate of not less than 0.5°C per hour. Such a capacity requires 1 tonne (3.5 kW) of refrigeration capacity per 18 tonnes of produce for small stores of up to about 150 tonnes capacity, and about 1 tonne per 25 tonnes for larger stores. The capacity for larger stores can be varied by having two or more compressors or by the technique of cylinder unloading in one compressor.

Good temperature control requires spatial variation of no more than ±1°C and a variation in time in any one position of no more than ±0.5°C. A temperature difference of 1°C over the storage period has significant effects on most produce, especially those stored at less than 5°C. The optimum thickness of insulation in walls and the ceiling is the equivalent to 4 cm of cork per 10°C difference in temperature between the inside and outside. This will keep the overall heat transfer to about 0.3 kJ/m²/h. This gives about the most economical ratio of the cost of refrigeration capacity to the cost of insulation, and also enables high humidities to be maintained. The best insulation is the cheapest that gives the required performance; it may be 6 cm thickness of poly-urethane foam or 40 cm of sawdust. Floors generally require half the thickness of insulation that is used in the walls. A vapour barrier of thick polyethylene, laminated foil material, or the equivalent, having a low water vapour transmission rate is placed on the warm side of the insulation to prevent moisture from migrating to and condensing within the insulation.

Cool stores may be constructed in many ways. As long as the above conditions are met, all can be satisfactory. Many modern cool rooms are either of sandwich panel construction with polystyrene foam slabs as the insulation in the prefabricated panels, or foamed-in-place polyurethane is applied to the inner faces of the structure. The 'skins' on the outsides of the insulation are metal, commonly either aluminium or zinc-coated steel or waterproof plywood (Figure 7.3). Floors are constructed of reinforced concrete capable of carrying point loads from fork lifts as well as stacking loads. Cooled air is generally supplied by forced or induced draft coolers (Figure 7.4), consisting of framed, closely spaced and finned evaporator coils fitted with fans to circulate the air over the coils. Some means of defrosting the coils is also required when storage temperatures are low and the coil surface operates at temperatures below 0°C.

**Figure 7.3**
Modern cool room being constructed of metal clad panels insulated with polystyrene insulation. A Colorbond finish has been applied to the panels to facilitate cleaning the cool room. All joints and entry points for electrical cables and plumbing are sealed with waterproof silicone mastic to prevent entry of water into the insulation.

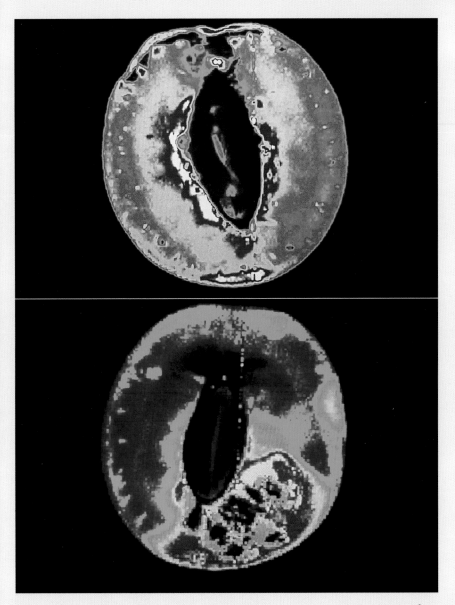

Non-destructive, colour-enhanced, proton magnetic resonance, transverse section images of ripened 'Kensington Pride' mango fruit. The darker areas have lower water activity and lighter areas have higher water activity. TOP: Fruit damaged by a heat disinfestation treatment applied when the fruit was mature green. The dark area in the centre of the fruit is the seed, and the dark areas just inside the periphery of the fruit show heat-injured parts of the mesocarp, where starch-to-sugar hydrolysis during ripening was impaired. BOTTOM: Fruit infested by a Queensland fruit fly larva which developed from an egg oviposited when the fruit was mature green. The dark area in the centre of the fruit is the seed, and the dark areas bounded by red areas, between the periphery of the fruit and the seed, show parts of the mesocarp where air-filled galleries arose as a result of the larva feeding on the tissue.

**Figure 8.4**

Illustrations of examples of non-pathogenic diseases:

(A) occurrence of chilling injury of sweet potato following 7 weeks storage at (left to right) 0°C, 5°C, 10°C and 15°C;

(B) bitter pit of apple;

(C) core flush of apple stored at 0°C;

(D) storage spot of Valencia oranges stored at 5°C;

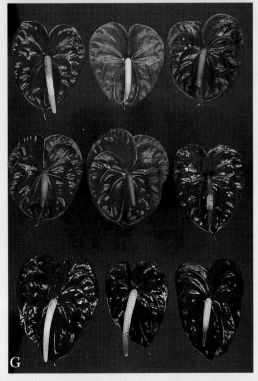

(E) blackheart of pineapple (in Australia this condition is usually caused by preharvest chilling);

(F) superficial scald of Granny Smith apple stored at 0°C;

(G) chilling injury in anthurium stored at 0°C (top row) and 5°C (middle row) for 3 days, followed by 2 days at 20°C (the flower in the bottom row was held for 3 days at 15°C, followed by 2 days at 20°C);

(H) chilling injury of frangipani (top), following 3 days at 4°C and 3 days at 20°C (the non-chilled flower [bottom] was held for 6 days at 22°C).

(Courtesy of Mr Brian Beattie, NSW Department of Agriculture [plates A–E], Mr John Golding, University of Western Sydney Hawkesbury [plate F] and Dr D.H. Simons, The University of Queensland [plates G–H]).

H

H

**Figure 9.1**
Illustrations of pathological diseases:
(A) brown rot of peach (*Monilinia fructicola*);

(B) blue mould *Penicillium italicum* (left) and green mould *Penicillium digitatum* (right) of oranges;

(C) anthracnose (*Colletotrichum gloeosporioides*) of avocado;

(D) crown rot of bananas caused by several species of fungi;

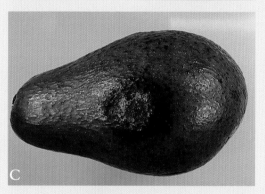

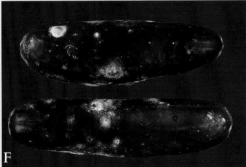

(E) stem-end rot of mango (*Dothiorella dominicana* is the most common cause of this rot in Australia);

(F) secondary fungal decay of cucumber at 20°C following chilling at 0°C.

(Courtesy of Brian Beattie, NSW Department of Agriculture [plates A–E] and Dr D.H. Simons, The University of Queensland [plate F]).

**Figure 9.4**
A non-pathogenic fungus, *Paecilomyces* sp. (dark fungus inoculated at the centre of the agar plate), engulfing *Geotrichum candidum*. (Courtesy of Brian Wild, NSW Dep. of Agr.)

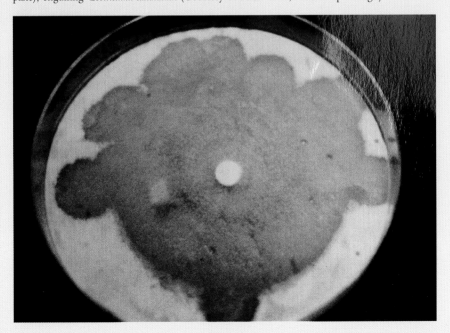

**Figure 10.2**

Ripening scale of 'Cavendish' banana (*Musa acuminata*, var Williams). The plates show the changes in colour of a single banana at 20°C at daily intervals: Day 0, initial colour before application of ethylene at 100 μL/L for 24 hours; Day 1, colour immediately following treatment with ethylene; Days 2–9, the progressive yellowing of the fruit. Soluble solids concentration reached a maximum in comparable fruit at about day 8. Ripe fruit spots did not appear until day 9.

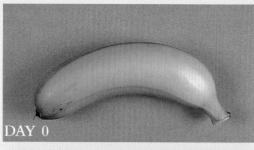
DAY 0

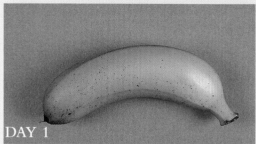
DAY 1

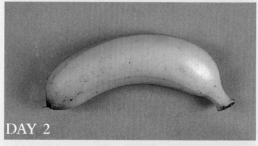
DAY 2

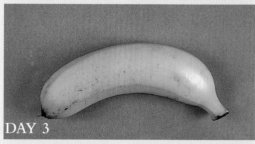
DAY 3

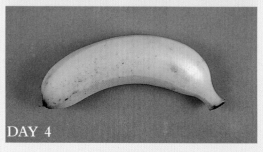
DAY 4

DAY 5

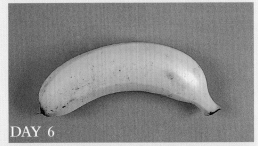

DAY 6

DAY 7

DAY 8

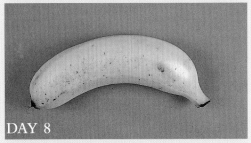

DAY 9

## STAGE 1

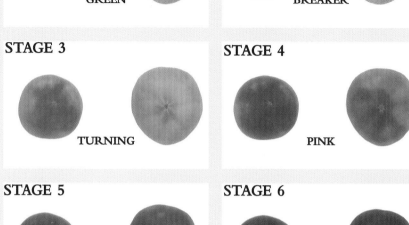

GREEN

## STAGE 2

BREAKER

## STAGE 3

TURNING

## STAGE 4

PINK

## STAGE 5

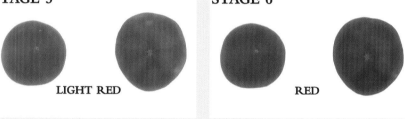

LIGHT RED

## STAGE 6

RED

**Figure 10.4**
Tomato colour chart. Individual fruit ripened
at 20°C were photographed daily. Stage 2,
day 2 of the climacteric rise in respiration and
ethylene production coincides with colour
stage 2 (breaker). Stage 6, tomatoes reach an
overall red stage 4 days from the breaker
stage. At least 2 more days of ripening are
required for the fruit to develop full flavour.
SOURCE W.B. McGlasson, B.B. Beattie and
E.E. Kavanagh (1985) Tomato ripening
guide, *Agfact,* H8.4.5, NSW Department of
Agriculture.

## STAGE 7

RED RIPE

**Figure 7.4**
A modern cool room in which the cool air is blown through the coils by fans mounted at the back of the coils (forced draft cooler). The more common arrangement is to mount the fans in front of the coils (induced draft cooler). IDCs are preferred because they give better air distribution. In this example the palletised produce is placed on racks.

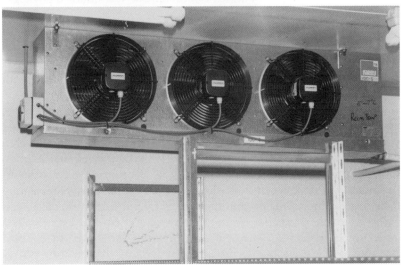

# DESIGN AND CONSTRUCTION OF CONTROLLED ATMOSPHERE STORES

Controlled atmosphere storage of apples and pears was initially a system of ventilated gas storage, in which the atmosphere was generated by the accumulation of carbon dioxide from fruit respiration, and the level of carbon dioxide was maintained at the desired level by ventilation with outside air. Atmospheres of such stores typically contained 5–10 per cent carbon dioxide and 6–11 per cent oxygen as one volume of carbon dioxide is produced by the fruit for each volume of oxygen consumed. Further research revealed that some important cultivars were damaged by carbon dioxide concentrations above 3 per cent and that significant benefits could be obtained if store oxygen concentrations were in the range of 2–3 per cent. An atmosphere of 2–3 per cent carbon dioxide with 2–3 per cent oxygen was, therefore, found to be suitable for many apple and pear cultivars at cool storage temperatures (Chapter 6). To be able to maintain such a low oxygen atmosphere, a much more gas-tight room was required. This required highly specialised methods of construction. Furthermore, ventilation of the store with outside air to control the carbon dioxide concentration was not possible as it would introduce too much oxygen; therefore, some means of absorbing or 'scrubbing out' the excess carbon dioxide was required. Early carbon dioxide scrubbers relied on chemical absorption of the carbon dioxide in alkaline solutions, such as potassium hydroxide or calcium hydroxide. Later, less cumbersome and less messy methods were developed by which carbon dioxide was adsorbed physically or absorbed chemically with dry hydrated lime.

Thus the controlled atmosphere store has to be relatively gas-tight, and fitted with a means of measuring and controlling the concentrations of both carbon dioxide and oxygen. Being a sealed chamber, the refrigeration system has to be completely reliable, and the room has to be fitted with adequate, accurate and reliable remote-reading thermometers.

In the early 1970s an external generator was developed that consumed oxygen in the air much more rapidly than the produce could consume by respiration. The generator operated on gaseous fuel, which either produced a low oxygen atmosphere with the required carbon dioxide content to flush out the oxygen from the room (flushing system) or consumed the oxygen from the air in the room itself (recirculating system). A carbon dioxide absorber was required to absorb the excess carbon dioxide produced by the generator and by the fruit. Such a generator enabled a 2–3 per cent oxygen atmosphere to be maintained in a relatively leaky room. In modern practice, low oxygen atmospheres can be achieved by flushing the room with liquid or compressed nitrogen in locations where nitrogen is relatively cheap. External generators are being superseded by gas separators such as the pressure swing adsorption (Figure 7.5) or hollow fibre membrane systems (Figure 7.6), which can generate gas streams containing low oxygen concentrations by separating oxygen and nitrogen from air. These separators have the major advantage of not producing any undesirable products arising from the incomplete combustion of fuels. Oxygen concentrations can be reduced rapidly by flushing the cool room. Once the required oxygen levels are reached, carbon dioxide concentrations are controlled by scrubbing with activated charcoal adsorbers. These modern carbon dioxide scrubbers have largely replaced the more cumbersome system of using hydrated lime in paper sacks. The operation of low-oxygen stores is now relatively simple and is frequently automated. Despite these advances, it is imperative to build gas-tight stores to ensure that separators are used economically. These land-based systems for generating and maintaining controlled atmospheres have now been adapted to ship refrigerated holds and to refrigerated shipping containers.

**Figure 7.5**

Pressure swing adsorption machine. Filtered dry compressed air is passed alternatively through chambers packed with a molecular sieve that retains oxygen when under high pressure and allows compressed nitrogen of at least 99 per cent purity to be generated. When the adsorption capacity in one chamber is nearing completion, the compressed air is automatically switched to the second chamber and the first is purged at atmospheric pressure to release the adsorbed oxygen.

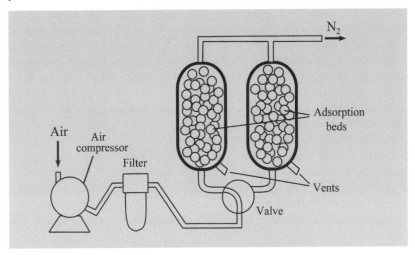

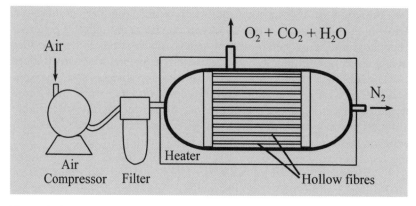

**Figure 7.6**
Hollow fibre membrane system. High pressure, dry, oil-free air is passed through hollow fibres constructed from a polymer that is differentially permeable to gases. Oxygen, carbon dioxide and ethylene permeate the fibres rapidly and are purged to the atmosphere, enabling the generation of a high pressure stream containing up to 99.5 per cent nitrogen.

## Construction

An essential feature of a controlled atmosphere store is the provision of an effective gas barrier, which is most conveniently placed directly on the inside of the insulated surface. If the external vapour barrier is defective, however, moisture that penetrates this barrier will then be contained on the inside of the gas barrier, leading to water-logging and destruction of the insulation. During the early days of controlled atmosphere storage of apple and pears, existing cool rooms were frequently converted. The range of sealants was limited and it was frequently difficult to achieve sufficient gas tightness. The concept of the 'jacketed' room overcame these problems and allowed the gas barrier to remain readily accessible. The system provides a gas-tight lining inside a cool room, with an air space between the insulated walls and the lining. Cold air is circulated through this narrow space to remove the heat. An expensive disadvantage of the system is the need for air ducts under the floor. The 'blanket' type of store is a variation of the jacket type, and has a normal floor that reduces the cost of construction. Primary cooling air circulates over the ceiling and around the walls. A further variation is a room in which a plastic film acts as a blanket (i.e. 'plastic tent' controlled atmosphere storage). The tent is flexible and automatically varies the internal volume as pressure differentials develop, thus avoiding significant pressure effects.

Jacketed systems and plastic tents have been superseded with the advent of foamed-in-place polyurethane. The use of polyurethane enables the satisfactory conversion of existing cool stores to controlled atmosphere operation, and also provides a cheap new method of construction by completely lining the interior of a simple metal chamber. If correctly applied, the polyurethane provides both insulation and a gas barrier. A pressure relief device (burp valve), usually a water trap, is fitted through the walls of such a rigid, gas-tight structure to avoid damage by limiting pressure differentials to 370 Pa (15 mm water gauge).

### Controlled atmosphere stores are lethal

While controlled atmosphere stores support plant life at a low level, they will not support mammalian life. Controlled atmosphere stores should be treated with respect to ensure that no one is ever exposed to such an atmosphere, unless wearing an efficient respirator with its own oxygen supply. Transport vehicles, in which the atmosphere has been modified with liquid nitrogen or modern controlled atmosphere systems, are especially dangerous and, like controlled atmosphere stores, must be well ventilated before entry.

## MANAGEMENT OF PRODUCE STORAGE

High quality produce will come out of storage only if it is of high quality on entering the store, and if management of the storage facilities is of high standard. Given correct selection and handling of the fruit, the success of subsequent storage depends on: quickly reducing the temperature of the fruit to the desired level and maintaining it with little variation; close maintenance of the desired humidity and gas concentrations in the storage atmosphere; and avoiding over-storage.

### Precooling the rooms

Storage rooms are generally brought down to the appropriate temperature a few days before fruit is to be stored. Three days is enough for a fully insulated room, but rooms without floor insulation should be precooled for a week to ensure that the floor has cooled to equilibrium. Failure to precool the room before loading is often the cause of unsatisfactory temperature maintenance, slow cooling, and excessive shrinkage of the produce.

### Temperature control

Air movement transfers heat from the fruit to the coils by natural convection in a room with overhead grids (cooling pipes); by forced circulation in rooms cooled by forced or induced draft coolers; or by a combination of natural and forced convection. It follows that the nature of the packages and the method of stacking packages must allow the air to move readily through all parts of the stack for the produce to be cooled quickly and uniformly.

Spatial variation in the temperature of produce in a good store should not exceed 1°C above or below the nominated storage temperature. Several factors influence the spatial distribution of temperature in a store. The most important single requirement for uniform produce temperature is uniform cooling over the entire area on the top of the stack. This applies equally to the distribution of air from forced air circulation systems and to the even distribution of the coils over the ceiling in natural circulation rooms. Also of importance is the uniformity of the air paths through the stow as air always takes the path of least resistance. Ideally there should be a continuous, narrow air slot in the direction of air flow past at least two faces of every box or carton and each side of every bulk bin, together with no large vertical gaps in the stack to allow short-circuiting by the cool air. The room should be well insulated to reduce heat leakage, and the coolers should have ample capacity to ensure a small difference between the temperature of the air and coil surface.

## Selection, sorting and handling of produce

It is desirable to sort and grade the size of produce before storage. Not all commodities are fit for storage; some have better keeping qualities than others, some are blemished (and usually sold to processors), and some of the harvested produce is unmarketable. Refrigerated storage is expensive, and it is not economical to have cool storage space occupied by produce that is not fit for sale or produce that would be better marketed immediately. Sorting and sizing before storage, so that both quantity and quality in storage are known, is of great assistance in orderly marketing.

## Loading

If possible, warm produce should be cooled in a separate cool room from that used for storage. If only one room is available, the designed daily intake (commonly 10 per cent of capacity) should not be exceeded. Otherwise, the life of the produce will be reduced and shrinkage promoted. Warm produce should be loosely stacked, and cooling can be improved with the aid of an auxiliary portable fan placed in front of the stack, with the suction side facing the produce to draw air through it (Chapter 4).

## *Stacking*

It is bad practice to overfill a room as this results in variable temperatures and therefore a poor outturn of a proportion of the produce. Packaged produce is carefully stacked to give economy of space, adequate and uniform air circulation, and accessibility. The following requirements for stacking are essential for rapid cooling and good temperature control for any type of package:

1.  Keep the stack 8 cm away from outer walls and 10–12 cm away from any wall exposed to the sun. This will ensure that heat coming in through the walls will be carried away to the coils by air moving freely between the stack and the wall without warming nearby produce.

2.  Leave a clear air space of not less than 20 cm between overhead coil drip trays and the top of the stack. If unit coolers or other forced air circulation systems are used, the clear space between the top of the stack and the ceiling is generally not to be less than 25 cm. This ensures that a uniform layer of cold air blankets the whole stack. The full depth of the space in front of a forced draft cooler is kept clear for a distance of 2 metres to allow it to function properly and to avoid freezing produce.

3.  An air plenum of about 8 cm is required between the floor and the stack. When bins and pallets of boxes are used, the pallet bases provide the necessary air gap above the floor. Wherever possible, they are placed with the pallet bases parallel to the direction of airflow (i.e. running towards the forced draft cooler).

4.  Leave small, vertical air paths within the stack, not less than 1 cm wide between adjacent packages. Freely exposed cartons cool at a similar rate to that of packed boxes of the same dimensions. Cartons, having straight sides, require special treatment in stacks. A layer-reversed, open chimney stacking pattern will provide the necessary vertical gaps between cartons (1 cm) and at the same time provide a stable stack.

5.  Bulk storage bins should have air gaps in the floor of 8–10 per cent of the base area. Rapid cooling of produce is possible in such bins. Bins of warm produce are first stacked only two-high overnight to allow quick removal of field heat from the produce. Next day they may be stacked to full height. Unless a high RH is maintained in the cool store, produce in bins that also have air gaps in the side may shrivel excessively. The sides of the bins can be lined to reduce shrinkage, but cooling will be slow if the slatted bottoms are lined. Vertical air gaps about 4 cm wide are left around each column of bins — at least at the sides, which are at right angles to the pallet-base bearers — as this allows free escape of the air rising by convection through the produce in the bin.

## Weight loss and shrinkage

Excessive shrinkage is due to: immaturity of the produce; delay before storage; picking produce when hot and then placing hot produce in the cool store; packing produce into dry wooden boxes or cartons; high storage temperatures, including hot spots in the room; low humidities due to insufficient insulation or insufficient coil surface; slow cooling; and excessive air circulation. Fast cooling, uniformly low temperatures and high humidities in the store are therefore necessary for low weight loss. The extra cost of additional cooling and insulation, and a good vapour barrier, can be more than offset by reduced weight loss and better produce condition after storage.

Weight loss during cooling may also be greatly reduced by wetting warm produce, such as leafy vegetables, before it is put into the store. It is preferable to harvest produce early in the morning, when it is coolest, and to put it directly into the cool store. This reduces the load on the refrigeration plant and lowers costs. When it is necessary to harvest produce later in the day during hot, dry weather, it may be practicable to spray some types of produce with water, to leave produce overnight in the open to cool by a combination of evaporative cooling and radiation cooling (if the night sky is clear), and to place produce into store the next morning.

## Orderly marketing and over-storage

Over-storage is still one of the most common faults in the cool storage of produce. It is good marketing practice to commence selling long-keeping produce, such as apples and pears, from the cool store early and to continue regularly throughout the season. To achieve orderly marketing, produce in the cool store needs to be segregated according to the expected keeping quality and removed for sale accordingly; as a general rule produce first in should be first out.

Over-storage of produce may be minimised by placing aside a few small units of the various lines of produce; these are removed to room temperature at intervals during storage, and the whole line should be marketed without delay at the first sign that the sample has deteriorated. A further important reason for making such regular inspection of samples during storage is that some fruit may look in fine condition in the cool store but may either develop physiological disorders or fail to ripen satisfactorily after removal to ambient temperature.

## Sanitation

Cool rooms should be thoroughly cleaned at the end of each season and, if necessary, sterilised to reduce the risk of losses by mould attack. The walls and floor can be washed with a solution of sodium hypochlorite (chlorine) followed by fumigation with formaldehyde gas. Mouldy or otherwise contaminated bins and boxes should be cleaned and sterilised with steam or a fungicide before reuse. Grading machines are often an important source of mould contamination that leads to the development of rots in storage. These machines should be cleaned and swabbed or sprayed with a fungicidal solution daily. The equipment should be regularly inspected for defects and any points likely to cause produce injury should be repaired.

## REFRIGERATED TRANSPORT

By land or by sea, much produce is transported long distances under refrigeration. Refrigerated road or rail vehicles can be regarded as insulated boxes fitted with modular mechanical refrigeration units powered from diesel units. Refrigerated ships have a central refrigerating plant; the whole ('reefer ships') or only part of the carrying space on a vessel may be insulated and refrigerated.

Much refrigerated sea freight is now carried in containers of 30 or 60 cubic metres capacity, which permit temperature control from door to door. One type of refrigerated freight container, the 'integral' container, has its own refrigeration unit, operated electrically, and perhaps also incorporates a diesel-powered generator (Figure 7.7). The other type, the 'porthole' container, is a passive unit that must be supplied with cool air from a clip-on refrigeration unit or from a central unit (Figure 7.8).

Economy of space is a prime requirement in all transport, and therefore refrigerated transport vehicles and containers are designed for high density stowage. They are not designed for rapid cooling, so successful refrigerated transport requires thorough precooling of the load. Respiratory heat is a significant proportion of the refrigeration load. Therefore, unless the journey is short, some air space must be provided between the packages during stowage of produce. Significant amounts of heat enter refrigerated road transport vehicles from solar radiation of the outside air, heat reflected from the road and from air leakage through the doors. To ensure the maintenance of even temperatures it is necessary to provide good circulation of cooling air around the load (Figure 7.9). Rules covering precooling, stowage and air circulation have been developed from research and commercial experience. Maximum acceptable loading temperatures are commonly specified and ought to be closely policed.

**Figure 7.7**
Integral-insulated shipping container with a built-in refrigeration unit.
**Figure 7.8**
A porthole refrigerated container must be coupled to either a clip-on refrigeration unit or to a source of cool air generated by a land-based or a shipboard refrigeration unit. In this example of a land-based system, the refrigeration plant can serve a battery of containers. When the containers are in place, the shutters are moved aside to allow connection to the air ducts.

**Figure 7.9**
To ensure effective air circulation in a refrigerated road vehicle there must be an air delivery chute, ribs on the doors and walls, channels or pallets on the floor and a return bulkhead.
SOURCE A.K. Sharp, A.R. Irving and A.A. Beattie (1985) Transporting fresh produce in refrigerated trucks, *Agfact*, H1.4.3, NSW Department of Agriculture, Sydney.

# FURTHER READING
·······

Anon (1989) *Guide to food transport: Fruit and vegetables*, Mercantile Publishers, Copenhagen.

American Society of Heating, Refrigerating and Air-Conditioning Engineers (1986) *ASHRAE handbook of refrigeration systems and applications*, Atlanta, GA.

Irving, A.R. (1988) *Code of practice for handling fresh fruit and vegetables in refrigerated shipping containers*, Department of Primary Industries and Energy, Canberra.

Kader, A.A. (1985) *Postharvest technology of horticultural crops*, University of California, Davis, CA. Special Publication no. 3311.

Story, A. and D.H. Simons (eds) (1997) *Fresh produce manual*, Australian United Fresh Fruit and Vegetable Association Ltd, Sydney.

Thompson, A.K. (1995) *Postharvest technology of fruit and vegetables*, Blackwell Science, London.

# 8
# PHYSIOLOGICAL DISORDERS

Physiological disorders refer to the breakdown of tissue that is not caused by either invasion by pathogens (disease-causing organisms) or by mechanical damage. They may develop in response to an adverse preharvest and/or postharvest environment, especially temperature, or to a nutritional deficiency during growth and development.

## LOW TEMPERATURE DISORDERS

Storage of produce at low temperature is beneficial because the rates of respiration and of general metabolism are reduced (Chapter 4). Low storage temperatures do not, however, suppress all aspects of metabolism to the same extent. Some reactions are sensitive to low temperature and cease completely below a critical temperature. Several such cold-labile enzyme systems have been isolated from plant tissues. Decreasing temperature does not reduce the activity of other systems to the same extent as it does respiration. For these systems, this differential response leads to an accumulation of reaction products and possibly a shortage of reactants, while the converse occurs with cold-labile systems. The overall effect is that an imbalance in metabolism is created. If the imbalance becomes serious enough so that an essential substrate is not provided or toxic products accumulate, the cells will cease to function properly and will eventually lose their integrity and structure. These collapsed cells manifest themselves as areas of brown tissue in the produce. Metabolic disturbances occurring at subambient temperature are generally divided into two main groups: chilling injury and physiological disorders.

### Chilling injury
Chilling injury is a disorder that has long been observed in plant tissues, especially those of tropical or subtropical origin. It results from the exposure of susceptible tissues to temperatures below about 15°C, although the critical temperature at which chilling injury symptoms are produced varies for different commodities. Chilling injury is a separate

phenomenon from freezing injury, which results from the freezing of the tissue and the formation of ice crystals at temperatures below the freezing point. A clear distinction can therefore be made between the causes of chilling and freezing injuries. Susceptibility to chilling injury and its manifestations vary widely among different commodities. In addition, commodities grown in different areas may behave differently, and varieties of the same crop can also behave quite differently in response to similar temperature conditions.

Table 8.1 summarises the physical symptoms of chilling injury and the lowest safe storage temperature for some fruits. A common chilling injury symptom is pitting of the skin, usually due to the collapse of the cells beneath the surface, and the pits are often discoloured. High rates of water loss from damaged areas may occur, which accentuates the extent of pitting. Browning of flesh tissues is also a common feature. Browning often first appears around the vascular (transport) strands in fruit, probably

### TABLE 8.1   CHILLING INJURY SYMPTOMS OF SOME FRUITS

| PRODUCE | LOWEST SAFE STORAGE TEMPERATURE (°C) | SYMPTOMS |
|---|---|---|
| Avocado | 5–12* | Pitting, browning of pulp and vascular strands |
| Banana | 12 | Brown streaking on skin |
| Cucumber | 7 | Dark coloured, water-soaked areas |
| Egg plant | 7 | Surface scald |
| Lemon | 10 | Pitting of flavedo, membrane staining, red blotches |
| Lime | 7 | Pitting |
| Mango | 5–12 | Dull skin, brown areas |
| Melon | 7–10 | Pitting, surface rots |
| Papaya | 7–15 | Pitting, water-soaked areas |
| Pineapple | 6–15 | Brown or black flesh |
| Tomato | 10–12 | Pitting, Alternaria rots |

* A range of temperature indicates variability between cultivars in their susceptibility to chilling injury.

**Figure 8.1**
Chilling injury in avocado fruit appears as browning of the mesocarp due to the breakdown of cell compartmentalisation and the action of polyphenol oxidases to produce tannins (top). The fruit were stored in air for 21 days at 5°C and then ripened at 20°C for 5 days. The bottom fruit remained free of symptoms after ripening in air at 20°C following storage for 21 days at 5°C in sealed polyethylene bags that generated atmospheres typically 7 per cent carbon dioxide, 3 per cent oxygen and 90 per cent nitrogen.
SOURCE K.J. Scott and G.R. Chaplin (1978) Reduction of chilling injury in avocados stored in sealed polyethylene bags, *Trop. Agric. Trinidad* 55: 87–90. (With permission.)

as a result of the action of the enzyme polyphenol oxidase on phenolic compounds released from the vacuole after chilling, although this has not been proved in all cases. Figure 8.1 shows typical browning symptoms of avocado fruit affected by chilling injury. Fruit that has been picked immature will fail to ripen or will ripen unevenly or slowly after chilling. Degreening of citrus is slowed by even mild chilling. Water-soaking of leafy vegetables and some fruits such as papaya is also often observed. The symptoms of chilling injury normally occur while the produce is at low temperature, but sometimes will only appear when the produce is removed from the chilling temperature to a higher temperature. Deterioration may then be quite rapid, often within a few hours.

Chilling injury causes the release of metabolites, such as amino acids and sugars, and mineral salts from cells that together with the degradation of cell structure provide an excellent substrate for the growth of pathogenic organisms, especially fungi. These pathogens are often present as latent infections or may contaminate the produce either during harvest or postharvest during transport and marketing. For this reason, increased rotting is a common occurrence in tropical produce

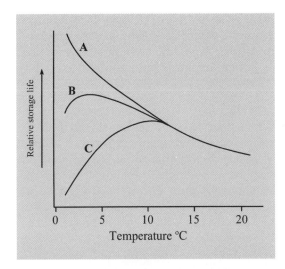

**Figure 8.2**
Storage life at various temperatures of produce with no (A), slight (B) or high (C) sensitivity to chilling injury.
SOURCE R.G. Tomkins, *The choice of conditions for the storage of fruits and vegetables,* East Malling Research Station, Ditton Laboratory Memoir no. 91. (With permission.)

after low temperature storage. Another consequence of chilling is the development of off-flavours or odours. The complex array of symptoms suggests that several factors are responsible for the development of chilling injury.

The temperatures quoted in Table 8.1 refer to the limiting or critical temperature below which some physical symptom of chilling injury will usually be observed. If the temperature is just below this critical temperature, relatively long exposure to the temperature will be required before injury is observed. Injury will generally appear more quickly and will be more severe the further the temperature is below the critical chilling temperature. Storage of the commodity may be possible for a useful period of time at temperatures slightly below the critical temperature where there is only a slight susceptibility to chilling injury. This relation between storage life at various temperatures and sensitivity to chilling is illustrated in Figure 8.2. Such a relationship also holds for the development of physiological disorders. These disorders are therefore subject to a time/temperature relationship.

The most obvious method for the control of chilling injury is to determine the critical temperature for its development in a particular fruit and then not expose the commodity to temperatures below that critical temperature. But exposure for only a short period to chilling temperatures with subsequent storage at higher temperatures may prevent the development of injury. This conditioning treatment has been found to be effective for preventing black heart in pineapple, woolliness in peach, and flesh browning in plum, but it is not known whether other produce will respond similarly. It has been claimed that modified atmosphere storage can reduce the extent of chilling in some produce. Also, maintenance of high relative humidity, both in storage at low temperature and after storage, may minimise pitting. More research is needed to confirm these claims.

## Mechanism of chilling injury

The events leading to chilling injury can be separated into primary events, by which the plant cells sense the lowered temperature, and the long-term responses or secondary events that ultimately lead to the death of the cells. The primary events are more or less instantaneous and are reversible, at least for a period of time. The secondary events are eventually irreversible and are manifest as the various necrotic and other symptoms of chilling injury. This concept is illustrated in Figure 8.3.

**Figure 8.3**
Time sequence of events leading to chilling injury
SOURCE G.R. Chaplin, personal communication.

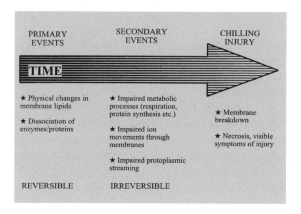

The critical temperature below which chilling injury will occur is characteristic of the species of plant and the commodity from which it is derived. Highly chilling-sensitive plants, such as bananas and pineapples, have a relatively high critical temperature of around 12°C or higher. It has even been suggested that for some pineapple cultivars the critical temperature may be greater than 20°C. Chilling-insensitive plants such as apples and pears have much lower critical temperatures of around 0°C or below. Of course, storage below about -1°C is not possible for fresh produce because of freezing damage.

The two most likely causes of chilling sensitivity are:

◆ a low temperature-induced change in the physical properties of cell membranes due to changes in the physical state of membrane lipids (the so-called lipid hypothesis of chilling); and

♦ low temperature-induced dissociation of enzymes and other proteins into their structural subunits, resulting in changes in the kinetics of enzyme activity and changes in structural proteins such as tubulin.

The lipid hypothesis is supported by data obtained using a number of physical techniques, including differential scanning calorimetry, electron spin resonance spectrometry and fluorescence polarisation of molecular probes intercalated into plant cell membranes. Data obtained by these techniques show that there is a change in the physical properties of extracted membrane lipids at characteristic temperatures in the range 7–15°C. The characteristic temperatures were found to coincide with the critical temperatures below which particular plant tissues or fruits showed symptoms of chilling injury. Only a very small proportion (<10 per cent) of the membrane lipids undergo the physical change, which is probably a phase separation. A physical change in the membrane lipids with lowering of the temperature would cause changes in the properties of the membranes. For example, ion and metabolite movements would be affected as would activities of membrane-bound enzymes. These changes could, in turn, cause imbalances in metabolism, with eventual disruption of the various membranes leading to the breakdown of cellular compartmentalisation, death of the cells and the appearance of symptoms of chilling injury.

There is some evidence from studies of both animals and plants that several enzymes of cellular metabolism undergo dissociation at temperatures approaching 0°C. Some multimeric enzymes split into their component subunits, with a consequent loss of enzymic activity and change in some kinetic properties. Some enzymes of both respiratory and photosynthetic metabolism are affected. The consequences of such changes in the relative activities of some enzymes will be imbalances in metabolism that would eventually lead to the death of cells. The toxin hypothesis of chilling injury — in which there is an accumulation of toxic products of metabolism, such as acetaldehyde — could be explained by imbalances in metabolism. Structural proteins of the cell cytoskeleton, such as tubulin, are cold-labile and undergo dissociation at low temperatures. This could account for the effect of low temperature on protoplasmic streaming, which is especially sensitive in chilling-sensitive plants.

Further research is required to fully elucidate the mechanisms of chilling injury. If it should turn out that only a few proteins are involved in the synthesis of the key lipids and cold-labile enzymes of metabolism then it may be possible in future to genetically engineer these proteins to make plants less chilling-sensitive and therefore improve the low temperature storage of subtropical and tropical fruits and vegetables.

## TABLE 8.2  SOME PHYSIOLOGICAL DISORDERS OF APPLES

| DISORDER | SYMPTOMS |
| --- | --- |
| Superficial scald | Slightly sunken skin discolouration, may affect whole fruit |
| Sunburn scald | Brown to black colour on areas damaged by sunlight during growth |
| Senescent breakdown | Brown, mealy flesh; occurs with overmature, overstored fruit |
| Low temperature breakdown | Browning in cortex |
| Soft (or deep) scald | Soft, sunken, brown to black, sharply defined areas on the surface and extending a short distance into the flesh |
| Jonathan spot | Superficial spotting of lenticels; occurs at higher temperatures |
| Senescent blotch | Grey superficial blotches on overstored fruit |
| Core flush (brown core) | Browning within core line |
| Water core | Translucent areas in flesh; may brown in storage |
| Brown heart | Sharply defined brown areas in flesh; may develop cavities |

## *Physiological disorders*

Physiological disorders mainly affect deciduous tree fruits, such as apples, pears and stone fruits, and most citrus fruits. Most of these disorders affect discrete areas of tissue, whether the produce be fruits, vegetables or ornamentals. (Figure 8.4 in the colour plate section). Some disorders may affect the skin of the produce but leave the underlying flesh intact; others affect only certain areas of the flesh or the core region.

With most disorders, the metabolic events leading to a manifestation of the symptoms are not fully understood and are not understood at all for many disorders. The discovery of most disorders could be considered 'non-scientific'. In the past, cool store operators or shipping agents have held fruit at low temperatures and found that they developed a variety of browning conditions. These conditions were given descriptive names as there was no other way of classifying the disorders. These names are still the only classification used. The apple has been studied more intensely

## TABLE 8.3  SOME PHYSIOLOGICAL DISORDERS OF FRUITS OTHER THAN APPLES

| PRODUCE | DISORDER | SYMPTOMS |
| --- | --- | --- |
| Pear | Core breakdown | Brown, mushy core in overstored fruit |
| | Neck breakdown, vascular breakdown | Brown to black discolouration of vascular tissue connecting stem to core |
| | Superficial scald | Grey to brown skin speckles; occurs early in storage |
| | Overstorage scald | Brown areas on skin in overstored fruit |
| | Brown heart | Same as for apple |
| Grape | Storage scald | Brown skin discolouration of white grape varieties |
| Citrus | Storage spot | Brown sunken spots on surfaces |
| | Cold scald | Superficial grey to brown patches |
| | Flavocellosis | Bleaching of rind; susceptible to fungal attack |
| | Stem-end browning | Browning of shrivelled areas around stem-end |
| Peach | Woolliness | Red to brown, dry areas in flesh |
| Plum | Cold storage | Brown, gelatinous areas on skin and flesh breakdown |

than other commodities and also appears to have the greatest variety of physiological disorders. Table 8.2 lists some of these disorders and their symptoms. These disorders require low temperature storage, usually at less than 5°C, for the development of symptoms. Each disorder is therefore presumed to be derived by a different metabolic route, although this may not prove to be true when their biochemistry is elucidated. Disorders in other fruits are shown in Table 8.3. When more research effort is devoted to commodities other than the apple, the list of physiological disorders will undoubtedly increase. There is no reason to believe that the apple should be more prone to develop disorders than other commodities.

Early studies of disorders found that, although a particular variety may be susceptible to a certain disorder, not all fruits develop the disorder.

**Figure 8.5**
Equipment for the flood application of diphenylamine to bulk boxes of apples and pears before cool storage. Boxes stacked two high are conveyed through the applicator. Supplementary jets on the side walls apply solution to the bottom boxes. A fungicide and a low concentration of calcium chloride may be added to the solution.

Susceptibility to disorders was shown to depend on a number of factors, such as maturity at harvest, cultural practices, climate during the growing season, produce size, and harvesting practices. The risk of a fruit developing a particular disorder can therefore be minimised by identifying susceptible fruits and not storing them for long periods. However, the market often has requirements that result in a preference for the type of fruit that is highly susceptible to a disorder. For example, consumers prefer large Jonathan apples with intense red colouration, but such fruits are susceptible to low temperature breakdown. Thus methods had to be developed to enable susceptible produce to be stored to meet consumer demand.

Various systems of temperature modulation have been developed to minimise the development of some disorders. Lowering the temperature in steps from 3°C down to 0°C in the first month of storage effectively minimises the development of low temperature breakdown and soft scald in apples. Low temperature breakdown of apples and stone fruits can also be reduced by raising the temperature to about 20°C for a few days in the middle of the storage period and then returning the fruit to low temperature. Such methods have not been widely adopted in commercial practice because of the logistical problems of having a whole room of produce ready to treat at one time and the difficulty of rapidly changing the temperature of a room full of fruit. A further problem is that any increase in the storage temperature will increase respiration, and thus shorten the storage life of produce held in the same room that is not susceptible to the disorder.

Controlled atmosphere storage can completely prevent Jonathan spot when as little as 2 per cent carbon dioxide is present. The incidence of core flush and various forms of flesh breakdown in apples is also often reduced in controlled atmosphere storage. However, in some instances the level of breakdown has reportedly increased in controlled atmosphere storage. This increase has been attributed to factors associated with controlled atmosphere storage other than the composition of the atmosphere. The enclosed room that is required for controlled atmosphere storage results in high humidity, restricted ventilation rates, and the accumulation of fruit volatiles in the atmosphere. These conditions are conducive to the development of apple breakdown. Superficial scald is another disorder that is enhanced in controlled atmosphere storage by these conditions (see below). Controlled atmosphere storage can also create new disorders if the produce is exposed to very high levels of carbon dioxide or low levels of oxygen for prolonged periods. The critical level of carbon dioxide that induces brown heart in apples and pears varies among different varieties and may be as low as 1 per cent. In addition to browning of the tissue, low oxygen injury is characterised by the development of alcoholic off-flavours produced by anaerobic metabolism.

The ultimate method for the prevention of a disorder is to understand the metabolic sequence that leads to the development of the disorder and then to prevent that metabolism from occurring. Chemical control is an obvious measure to prevent the development of disorders, but it is not necessarily the sole method possible. Storage disorders may also be minimised by physical and cultural treatments and by breeding less susceptible cultivars.

Skin blemishes are generally the more serious problem as even quite small skin marks render the fruit unacceptable in many markets. Internal defects can be tolerated to a greater extent as the consumer buys on visual inspection of the external appearance, and even upon consumption may never be aware of a small amount of internal browning. The skin disorders bitter pit and superficial scald of apple have received considerable attention, and control measures have been developed for both disorders (see Chapter 11).

Much is known about the metabolism relating to superficial scald. Early studies (before 1930) led to the hypothesis that superficial scald was caused by a toxic, volatile organic compound that accumulated in the apple during cool storage. In the 1960s, workers in Australia isolated $\alpha$-farnesene, a C15 sesquiterpene hydrocarbon from susceptible apple varieties, and suggested that it was the precursor compound involved in superficial scald. Being a fat-soluble compound, it accumulates in the lipid fraction on the skin. Oxidation products of $\alpha$-farnesene have been claimed to lead to the collapse of the cells and to tissue browning. Control of the disorder is achieved commercially by the application of various synthetic antioxidants, such as diphenylamine and ethoxyquin, which protect $\alpha$-farnesene against oxidation (Figure 8.5). Chilling injury is usually thought to develop along a different metabolic route to that for superficial scald, but $\alpha$-farnesene has been shown to accumulate in bananas during the development of chilling symptoms, suggesting that there may be a metabolic similarity in the two disorders. Recent studies in Australia has shown that superficial scald can be controlled by ethanol vapour in the atmosphere surrounding Granny Smith apples. While there may be some resistance to the use of ethanol on religious or liquor-related grounds, it raises the question of whether the use of ethanol vapour is a natural substance as it is metabolised, albeit in small quantities, by apples.

If satisfactory control methods are not available, the ultimate method of avoiding any physiological disorder is to hold susceptible fruits at a temperature high enough to prevent the disorder from being a problem. This temperature is usually 3–5°C, but is sometimes greater that 5°C. This partially negates the idea of using low temperature to minimise respiration, but it is better to market overmature produce than to have a disfiguring disorder present.

# MINERAL DEFICIENCY DISORDERS

Fruit and vegetables often show various browning symptoms that have been attributed to deficiencies in some mineral constituents of the produce. These disorders are prevented by the addition of the specified mineral, either during growth or postharvest, even though the actual role of the mineral in preventing the disorder has not been established for most disorders. Plants require a balanced mineral intake for proper development, so a deficiency in any essential mineral will lead to poor development of the plant as a whole. It can be said that the condition is a physiological disorder if the fruiting organ or actual 'vegetable' portion is affected rather than the whole plant.

## TABLE 8.4 CALCIUM-RELATED DISORDERS OF FRUIT AND VEGETABLES

| PRODUCE | DISORDER |
| --- | --- |
| Apple | Bitter pit, lenticel blotch, cork spot, lenticel breakdown, cracking, low temperature breakdown, internal breakdown, senescent breakdown, Jonathan spot and water core |
| Avocado | End spot |
| Bean | Hypocotyl necrosis |
| Brussels sprout | Internal browning |
| Cabbage | Internal tipburn |
| Chinese cabbage | Internal tipburn |
| Carrot | Cavity spot, cracking |
| Celery | Blackheart |
| Cherry | Cracking |
| Chicory | Blackheart, tipburn |
| Escarole | Brownheart, tipburn |
| Lettuce | Tipburn |
| Mango | Soft nose |
| Parsnip | Cavity spot |
| Pear | Cork spot |
| Pepper | Blossom-end rot |
| Potato | Sprout failure, tipburn |
| Strawberry | Leaf tipburn |
| Tomato | Blossom-end rot, blackseed, cracking |
| Watermelon | Blossom-end rot |

## Calcium

Calcium has been associated with more deficiency disorders than other minerals, and some examples are shown in Table 8.4. Some of these disorders, such as blossom-end rot of tomatoes, can be readily eliminated by the application of calcium salts as a preharvest spray while for others, such as bitter pit of apples, only partial control is obtained. However, this variability in the degree of control is probably related to the amount of calcium taken up by the fruit. For example, the use of postharvest dipping at subatmospheric pressures, which markedly increases the uptake of calcium, usually results in the total elimination of bitter pit.

A substantial amount of the added calcium binds with pectic substances in the middle lamella, and with membranes generally, and may prevent disorders by strengthening structural components of the cell without alleviating the original causes of the disorder. The strengthening of cell components may prevent or delay the loss of cell compartmentalisation and the enzyme reactions that cause browning symptoms.

Calcium has been found to be relocated in apples during storage. This

**141**

raises the possibility that a local deficiency can be created in one part of the tissue during storage, resulting in a manifestation of a physiological disorder in that region. Calcium has been shown to affect the activity of many enzyme systems and metabolic sequences in plant tissues. The addition of calcium to intact fruit or fruit slices generally suppresses respiration, but the response is concentration-dependent. The activities of isolated pectic enzymes, pectinmethylesterase (PME), exopolygalacturonase (exo PG) and endopolygalacturonase (endo PG), have shown differential responses to calcium concentration. The activity of PME is initially increased by increasing concentrations of calcium but is inhibited at higher concentrations. The large form of endo PG (PG1) extracted from tomato fruit is slightly stimulated by concentrations of calcium that inhibit the smaller endo PG forms of the enzyme. Calcium is needed for the activity of exo PG, kinases and a range of other enzymes. The ability of calcium to regulate these various systems has led to speculation that calcium may have a role in the initiation of the normal fruit ripening process. It is also possible that calcium prevents or delays the appearance of some physiological disorders by maintaining normal metabolism.

## Other minerals

Boron deficiency in apples leads to a condition known as internal cork. This condition is marked by pitting of the flesh and is often indistinguishable from bitter pit. The differences between the two disorders are that internal cork is prevented by the application of boron sprays and develops only on the tree, while bitter pit responds to calcium treatment and can develop postharvest.

The major mineral in plants is potassium, and both high and low levels of potassium have been associated with abnormal metabolism. High levels of potassium have been associated with the development of bitter pit in apple so that both high potassium and low calcium levels are correlated with pit development. Low potassium is associated with changes in the ripening tomato and delays the development of a full red colour by inhibiting lycopene biosynthesis.

There may be roles for other minerals in the development of other disorders. Injections of copper, iron and cobalt have induced symptoms similar to low temperature breakdown and superficial scald in apples, but this does not necessarily mean they have a role in the development of the natural disorder. Heavy metals, especially copper, act as catalysts for the enzymic systems that lead to enzymic browning, the browning of cut or damaged tissues that are exposed to air. The levels of these metals are important in processed fruit and vegetables, whether they are derived from the produce or from metal impurities that are included during processing.

# FURTHER READING
......

Beattie, B.B., W.B. McGlasson and N.L. Wade (1989) Postharvest diseases of horticultural produce, vol. 1, *Temperate fruit*, CSIRO, East Melbourne.

Coates, L., T. Cooke, D. Persley, B. Beattie, N. Wade and R. Ridgway (1995) Postharvest diseases of horticultural produce, vol. 2, *Tropical fruit*, Department of Primary Industries, Queensland, Brisbane.

Ferguson, I.B. (1984) *Calcium in plant senescence and fruit ripening, Plant Cell Environ* 7: 477–89.

Graham, D. (1990) Chilling injury in plants and fruits: Some possible causes with means of amelioration by manipulation of post-harvest storage conditions, in S.K. Sinha, P.V. Sane, S.C. Bhargara and P.K. Agrawal (eds), *Proceedings of the International Congress of Plant Physiology*, Indian Agricultural Research Institute, New Delhi, pp. 1373–84.

Graham, D. and B.D. Patterson (1982) Responses of plants to low, nonfreezing temperatures: Proteins, metabolism and acclimation, *Annu. Rev. Plant Physiol.* 33: 347–72.

Patterson, B.D. and D. Graham (1987) Temperature and metabolism, in D.D. Davies (ed.), *The biochemistry of plants*, vol. 12, Academic Press, London, pp. 153–99.

Poovaiah, B.W. (1986) Role of calcium in prolonging storage life of fruits and vegetables, *Food Technol.* 40(5): 86–89.

Raison, J.K. (1985) Alterations in the physical properties and thermal responses of membrane lipids: Correlations with acclimation to chilling and high temperature, in J.B. St. John, E. Berlin and P.C. Jackson (eds) *Frontiers of membrane research in agriculture*, Beltsville Symposium 9, Rowman & Allanheld, Totowa, pp. 383–401.

Raison, J.K. and J.M. Lyons (1986) Chilling injury: A plea for uniform terminology, *Plant Cell Environ.* 9: 685–86.

Snowdon, A.L. (1990) A colour atlas of post-harvest diseases and disorders of fruits and vegetables, vol. 1, *General introduction and fruits*, Wolfe, London.

——(1991) A colour atlas of post-harvest diseases and disorders of fruits and vegetables, vol. 2, *Vegetables*, Wolfe, London.

Wang, C.Y. (ed.) (1990) *Chilling injury of horticultural crops*, CRC Press, Boca Raton, FL.

# 9
# PATHOLOGY

Wastage of horticultural commodities by microorganisms between harvest and consumption can be rapid and severe, particularly in tropical areas where high temperatures and high humidity favour rapid microbial growth. Furthermore, ethylene produced by rotting produce can cause premature ripening and senescence of other produce in the same storage and transport environment, and sound produce can be contaminated by rotting produce. Apart from actual losses due to wastage, further economic loss occurs if the market requirements necessitate sorting and repacking of partially contaminated consignments.

## MICROORGANISMS CAUSING POSTHARVEST WASTAGE
·······

Many bacteria and fungi can cause the postharvest decay. However, it is well established that the major postharvest losses of fruit and vegetables are caused by species of the fungi *Alternaria, Botrytis, Botryosphaeria, Colletotrichum, Diplodia, Monilinia, Penicillium, Phomopsis, Rhizopus* and *Sclerotinia* and of the bacteria *Erwinia* and *Pseudomonas* (Figure 9.1 in the colour plate section). Most of these organisms are weak pathogens in that they can only invade damaged produce. A few such as *Colletotrichum* are able to penetrate the skin of healthy produce. Often the relationship between the host (fruit or vegetable) and the pathogen is reasonably specific. For example, *Penicillium digitatum* rots only citrus and *P. expansum* rots apples and pears, but not citrus. Complete loss of the commodity occurs when one or a few pathogens invade and break down the tissues. This initial attack is rapidly followed by a broad spectrum of weak pathogens that magnify the damage caused by the primary pathogens. The appearance of many commodities may be marred by surface lesions caused by pathogenic organisms, without the internal tissues being affected.

## THE INFECTION PROCESS
·······

Fruit and vegetables are rotted by organisms that either infect the produce while still immature and attached to the plant or during the harvesting and subsequent handling and marketing operations. The infection process, particularly postharvest, is greatly aided by mechanical injuries to the skin of the produce, such as fingernail scratches and abrasions, rough handling, insect punctures and cut stems. Furthermore, the physiological condition of the produce, the temperature, and the formation of the periderm (see later in this chapter) significantly affect the infection process and the development of the infection. It is important to know the pattern of the infection process so that suitable treatment strategies may be developed to control or eliminate the infection.

## Preharvest infection

Preharvest infection of fruit and vegetables may occur through several avenues, such as direct penetration of the skin, infection through natural openings on the produce, and infection through damage. Several types of pathogenic fungi are able to initiate an infection on the surface of floral parts and on sound, developing fruit. The infection is then arrested and remains quiescent until after harvest, when the resistance of the host decreases and conditions become favourable for growth, such as when the fruit begins to ripen or the tissue senesces. Such 'latent' infections are important in the postharvest wastage of many tropical and subtropical fruits, such as anthracnose of mango and papaya, crown rot of banana and stem-end rots of citrus. For instance, spores of *Colletotrichum* germinate in moisture on the surface of the fruit, and the end of the germ tube swells within several hours of germination to form a structure known as an appressorium, which may or may not penetrate the skin before the infection is arrested. It can also apply to temperate produce, such as *Botrytis* infection of berry fruit and many cut flowers.

Weak parasitic fungi and bacteria may also gain access to immature fruit and vegetables through natural openings, such as stomata, lenticels and growth cracks. Again these infections may not develop until the host becomes less resistant to the invading organism, such as when the fruit ripens. It appears that sound fruit and vegetables can suppress the growth of these organisms for a considerable time, but little is known about the interaction of the invading microorganism and the host tissue. An example of this infection mechanism is the penetration of apple lenticels before harvest by spores of *Phlyctaena vagabunda*, which then manifest themselves in storage as rots around the lenticels.

## Postharvest infection

Many fungi that cause considerable wastage of produce are unable to penetrate the intact skin of produce, but readily invade via any break in the skin. The damage is often microscopic but is sufficient for pathogens present on the crop and in the packing house to gain access to the produce. In addition, the cut stem is a frequent point of entry for microorganisms, and stem-end rots are important forms of postharvest wastage of many fruits and vegetables. Postharvest infection can also occur through direct penetration of the skin by *Sclerotinia* and *Colletotrichum*, for instance.

### Factors affecting development of infection

Probably the most important factor affecting the development of postharvest wastage of produce is the surrounding environment. High temperature and high humidity favour the development of postharvest decay, and chilling injury generally predisposes tropical and subtropical produce to postharvest decay. In contrast, low temperature, low oxygen and high carbon dioxide levels and the correct humidity can restrict the rate of postharvest decay by either retarding the rate of ripening or senescence, depressing the growth of the pathogen, or both.

Many other factors affect the rate of development of an infection in fruit and vegetables. The host tissue — particularly the pH of the tissue — acts as a selective medium. Fruit generally has a pH below 4.5 and is largely attacked and rotted by fungi; many vegetables have a pH above 4.5 and consequently bacterial rots are much more common. Ripening fruit is more susceptible to wastage than immature fruit, so treatments that slow down the rate of ripening, such as low temperature, will also retard the growth of decay organisms. The underground storage organs, such as potato, cassava, yam and sweet potato, are capable of forming layers of specialised cells (wound-periderm) at the site of injury, thus restricting the development of postharvest decay. During commercial handling of potato, periderm formation is promoted by 10–14 days storage at 7–15°C and 95 per cent RH, a process known as curing. A type of curing process (possibly by desiccation) has been shown to reduce the wastage of orange by *P. digitatum*. When the fruit is held at high temperature (30°C) and humidity (90 per cent) for several days, the orange peel becomes less turgid and lignin is synthesised in the injured flavedo tissue.

# CONTROL OF POSTHARVEST WASTAGE

·······

## *Preharvest*

In most instances, control of postharvest wastage should commence before harvest in the field or orchard. Wherever possible, sources of infection should be eliminated, and sprays for the control or eradication of the causal organisms applied. Preharvest sprays are generally not as effective as postharvest application of the chemical directly to the commodity, although some of the systemic fungicides have shown good control of latent infections, such as lenticel rot of apples and brown rot of peaches. With some of the later, more specific fungicides, the development of resistance to the fungicide by the organism has occurred. For instance, the rapid development of resistance by *Penicillium* species to the benzimidazole group of fungicides strongly suggests that preharvest sprays of these fungicides would be unwise because of the opportunity for selection and growth of resistant strains, particularly if the same fungicide was being relied upon for postharvest control of *Penicillium*.

Careful handling during harvesting can minimise mechanical damage and reduce subsequent wastage due to microbial attack. Similarly, it is also unwise to harvest some fruits such as citrus after rain or heavy dew as the peel is turgid and easily damaged.

## *Postharvest*

Many physical and chemical treatments have been used for postharvest wastage control in fruit and vegetables. The effectiveness of a treatment depends on three main factors:

1. the ability of the treatment or agent to reach the pathogen;
2. the level and sensitivity of the infection; and
3. the sensitivity of the host produce.

The time of infection and the extent of development of the infection are critical in respect to whether it can be controlled. For example, *Penicillium* and *Rhizopus* invade wounds during harvest and subsequent handling operations and are much more easily controlled by fungicide application to the surface of the commodity than the grey mould of strawberry, which infects the fruit in the field some weeks before harvest or even at the time of flowering. It is recommended that fungicides be applied within 24 hours of harvest so that infections can be controlled before they become established.

*Physical treatments*

Postharvest wastage of produce may be controlled by low and high temperatures, modified atmospheres, correct humidity, ionising radiations, good sanitation and development of wound barriers. Low temperature handling and storage is the most important physical method of postharvest wastage control, and the remaining methods can be considered as supplements to low temperature. The extent to which low temperature and other environmental modifications can be used to control wastage depends on the tolerance of the tissue to that environment. For example, most tropical and subtropical produce is susceptible to chilling injury and therefore cannot be subjected to very low temperature.

Heat treatments in the form of either moist hot air or hot water dips have had some commercial application for the control of postharvest wastage in papaya, mango, stone fruits and cantaloupe. The advantage of hot water dipping is that it can control surface infections as well as infections that have penetrated the skin, and it leaves no chemical residues on the produce. The absence of chemical residues demands that recontamination of the produce by microorganisms be prevented by strict hygiene and possibly the application of a fungicide, although at lower levels than those required without a hot water dip. Hot water dips must be precisely administered as the range of temperature necessary to control wastage (50–55°C) approaches temperatures that damage produce. Ionising radiations are effective in inhibiting microbial growth, but can cause physiological damage and aberrant ripening (Chapter 11).

*Chemical treatments*

Chemical control of postharvest wastage has only become an integral part of the handling and successful marketing of fruit during the past 25 years, particularly in the development of the world trade in citrus, bananas and grapes. The level of control of wastage depends on the marketing strategy for the commodity and the type of infection. For citrus, which have a relatively long postharvest life, the aim of the treatment is to prevent primary infection and also sporulation so that nearby fruits are not contaminated. The strawberry has a short postharvest life, and treatment is aimed at preventing the spread of grey mould that infected the strawberries in the field. In other words, the treatment has to match the subsequent marketing of the commodity. There is no point in treating a short-life commodity with a fungicide that has a long residual activity. The success of a chemical treatment for wastage control depends on several factors:

1. the initial spore load;
2. the depth of the infection within the host tissues;
3. the growth rate of the infection;
4. the temperature and humidity; and
5. the depth to which the chemical can penetrate the host tissues.

Moreover, the applied chemical must not be phytotoxic (i.e. it must not injure the host tissues) and must fall within the ambit of the local food additive laws.

A wide range of chemicals has been, and still is, used for the control of postharvest wastage in fruit, particularly in citrus, banana,s grapes and strawberries. Table 9.1 lists some of these compounds, their common names, the pathogens against which they are effective and the fruit on which they are applied. The chemicals listed in Table 9.1 are generally fungistatic in action rather than fungicidal; that is, they inhibit spore germination or reduce the rate of germination and growth after germination rather than cause death of the organism, and must come into direct contact with the organism to be effective. A few chemicals, such as chlorine and sulphur dioxide ($SO_2$), are true fungicides. Chlorine is commonly added to wash-water to kill bacteria and fungi, and sulphur dioxide is lethal to *Botrytis* on grapes. Chemicals may be impregnated into wraps or box liners, or applied as fumigants, solutions and suspensions, or in wax.

*Development of chemical treatments for control of citrus wastage*

An instructive example of the development of postharvest chemical control is given for citrus. Green mould (*Penicillium digitatum*) and blue mould (*Penicillium italicum*) are the major postharvest diseases of citrus, with green mould being more prevalent in humid areas. The stem-end rots (Table 9.2) are also significant causes of postharvest losses in more humid climates. Borax and sodium carbonate were the first chemicals to provide a measure of wastage control; both of these treatments were largely superseded by the more effective compound, sodium *o*-phenylphenate (SOPP), in the 1950s. Before 1940, Tomkins in England had found that the undissociated form of SOPP, *o*-phenylphenol (HOPP), controlled citrus wastage, but it caused serious scalding of the skin. The advantage of SOPP was that it was not phytotoxic but was converted to the fungitoxic free phenol at the appropriate sites. The *o*-phenylphenate anion (OPP) diffuses selectively into injury sites and is converted to the undissociated form, preventing infection at these sites during storage or marketing. The SOPP-OPP system has broad spectrum activity and provides excellent control of *Penicillium* and the

**149**

## TABLE 9.1 CHEMICALS THAT HAVE BEEN USED AS POSTHARVEST FUNGICIDES

| NAME AND FORMULATION | PATHOGEN CONTROLLED |
|---|---|
| *Alkaline inorganic salts* | |
| sodium tetraborate (borax) | *Penicillium* |
| sodium carbonate | *Penicillium* |
| sodium hydroxide | *Penicillium* |
| *Ammonia and aliphatic amines* | |
| ammonia gas | *Penicillium, Diplodia, Rhizopus* |
| sec-butylamine | *Penicillium*, stem-end rots |
| *Aromatic amines* | |
| dicloran | *Rhizopus, Botrytis* |
| *Benzimidazoles* | |
| benomyl, thiabendazole, | *Penicillium* |
| thiophanate methyl | |
| carbendazim | *Colletotrichum*, other fungi |
| *Triazoles* | |
| imazalil | *Penicillium*, stem-end rots |
| prochloraz | *Penicillium* |
| guanidine | |
| guazitine | *Penicillium, Geotrichum* |
| *Hydrocarbons and derivatives* | |
| biphenyl | *Penicillium, Diplodia* |
| methyl chloroform | *Penicillium*, stem-end rots |
| *Oxidising substances* | |
| hypochlorous acid | Bacteria, fungi build-up in wash water |
| iodine | Bacteria, fungi |
| nitrogen trichloride | *Penicillium* |
| *Organic acids and aldehydes* | |
| dehydroacetic acid | *Botrytis* and other fungi |
| sorbic acid | *Alternaria, Cladosporium* |
| formaldehyde | Fungi |
| *Phenols* | |
| *o*-phenylphenol (HOPP) | *Penicillium* |
| sodium *o*-phenylphenate (SOPP) | *Penicillium*, bacteria, fungi |
| *Salicylanilide* | *Penicillium, Phomopsis, Nigrospora* |
| *Sulphur (inorganic)* | |
| sulphur dust | *Monilinia* |
| lime-sulphur | *Sclerotinia* |
| sulphur dioxide gas, bisulphate | *Botrytis* |
| *Sulphur (organic)* | |
| captan | Storage rots |
| thiram | *Cladosporium*, crown and stem-end rots |
| ziram | *Alternaria*, crown and stem-end rots |
| thiourea | *Penicillium* spores |
| thioacetamide | *Diplodia* |

SOURCES J.W. Eckert (1977) Control of postharvest diseases, in M.R. Siegel and H.D. Sisler (eds), *Antifungal compou* control, in W.F. Wardowski, S.C. Nagy, W. Grierson (eds), *Fresh citrus fruits*, AVI, Westport, CT, pp. 315–60.; J.M. Ogav Proceedings of the third international biodegradation symposium, 1975, Kingston, Applied Science, RI, London, pp.

**150**

| Host | Remarks |
|---|---|
| Citrus | Only reasonably effective; problem with B residues |
| Citrus | Only slightly effective |
| Citrus | Only slightly effective |
| Citrus | Good for fumigation of degreening and storage rooms |
| Citrus | Good control as dip or fumigant |
| Stone fruits, carrot, sweet potato | Very effective |
| Citrus | Effective at low concentration; resistance a problem; residue tolerance 0–10 µg/g |
| Banana, apple, pear, pineapple, stone fruit | |
| Citrus | Effective against benzimidazole-resistant strains and at low concentration |
| Citrus | Effective against benzimidazole-resistant strains |
| Citrus | Effective against benzimidazole-resistant strains |
| Citrus | Smell unpleasant |
| Citrus | Inhibits spore germination only |
| Produce | Good sterilant, no penetration of injury sites, corrosive to metal |
| Citrus, grapes | Staining problem, expensive |
| Tomato, citrus | Hydrolyses to hypochlorous acid |
| Strawberry | Dip not accepted by industry |
| Fig | Sterilant for picking boxes, storage rooms |
| Citrus | Causes fruit injury |
| Produce | pH control needed to prevent injury; residue tolerance 10–12 µg/g |
| Citrus, banana | Slight control |
| Peach | |
| Grapes | Sulphur dioxide gas needs moisture to be effective; inexpensive; no toxic residues |
| Various produce | |
| Strawberry, banana | |
| Banana | |
| Citrus | Toxic to man |

arcel Dekker, New York, pp. 269–352.; J.W. Eckert and G.E. Brown (1986) Postharvest citrus diseases and their and A.H. El Behadli (1975) Chemical control of postharvest diseases, in J.M. Sharply and A.M. Kaplan (eds), Wild (1975) Fungicides for postharvest wastage control in fruit marketing, *Food Technol. Aust.* 27: 477–78.

stem-end rots, but some resistance of *P. digitatum* to OPP has been reported where treated lemons have been stored for lengthy periods. A typical treatment for citrus consists of a 2 minute dip in a 2 per cent solution of SOPP tetrahydrate at temperatures up to 32°C and pH 11.7 or higher. SOPP may also be foamed onto the fruit or incorporated into either wax applied to the fruit as further protection after washing and/or an initial fungicide treatment.

The next major advance was the discovery of the fungistatic action of biphenyl against *Penicillium* and several other types of fruit wastage pathogens. Biphenyl has greatly assisted the development of world trade over the past 40 years. Biphenyl, impregnated into fruit wraps or into paper sheets placed in the fruit container, sublimes into the atmosphere surrounding the fruit, preventing sporulation of *Penicillium* on the surface of infected fruits and thus the transfer of the infection to neighbouring fruits. Biphenyl wraps are often used as a complementary treatment on export fruit that have been treated with SOPP. Some problems are associated with biphenyl treatment. The fruit have a characteristic hydrocarbon odour that tends to disappear within a few days after their removal from the wraps; residues on the fruit surface sometimes exceed the permissible limit, which varies from 70–100 µg/g, and strains of *Penicillium* and *Diplodia* have been able to develop resistance to biphenyl in areas where lemons were stored for up to 4 months. As a result, biphenyl is banned in some countries.

Although SOPP gives excellent control of citrus wastage organisms, the conditions of application — the maintenance of solution pH and strength and rinsing with water after SOPP application must be strictly adhered to in order to avoid rind injury — presented some difficulty for packing houses. Packers therefore welcomed the appearance of the benzimidazole group of fungicides 20 years ago. Thiabendazole (TBZ), benomyl, thiophanate methyl and carbendazim have a wide spectrum of antifungal activity at extremely low concentrations, but are inactive against *Rhizopus*, *Alternaria*, *Geotrichum* and soft rot bacteria. They are not phytotoxic and have low mammalian toxicity; in fact, TBZ was originally introduced as an anthelmintic (drench for worm control in stock) in 1961. Benzimidazoles show systemic activity; that is, the compounds may be distributed around the plant by the transpiration stream. Thiabendazole is absorbed and transported as TBZ, however benomyl and thiophanate methyl hydrolyse to carbendazim in water or in the plant and are translocated as carbendazim. It is believed that carbendazim provides most of the antifungal activity of these fungicides. The mode of action of these benzimidazole compounds appears to be

## TABLE 9.2 EXAMPLES OF MAJOR POSTHARVEST DISEASES OF FRESH FRUITS AND VEGETABLES

| CROP | DISEASE | PATHOGENS |
|---|---|---|
| Apple, pear | Lenticel rot | *Phlyctaena vagabunda* Desm. Arx (= *Gloeosporium album* Osterw.) |
| | Blue mould rot | *Penicillium expansum* (Lk.) Thom |
| Banana | Crown rot | *Colletotrichum musae* (Berk. and Curt.) Arx (= *Gloeosporium musarum* Cke. and Mass.) *Fusarium roseum* Link amend. Snyd. and Hans. *Verticillium theobromae* (Turc.) Hughes *Ceratocystis paradoxa* (Dade) Moreau (= *Thielaviopsis paradoxa* [de Seynes] Hohn.) |
| | Anthracnose | *Colletotrichum musae* (Berk. and Curt.) Arx (= *Gloeosporium musarum* Cke. and Mass.) |
| Citrus fruit | Stem-end rot | *Phomopsis citri* Faw. *Diplodia natalensis* P. Evans *Alternaria citri* Ell. and Pierce |
| | Green mould rot | *Penicillium digitatum* Sacc. |
| | Blue mould rot | *Penicillium italicum* Wehmer |
| | Whisker mould | *Penicillium ulaiense* Hsieh, Su and Tzean |
| | Sour rot | *Geotrichum candidum* Lk ex Pers. |
| Grape, apple, pear, strawberry, leafy vegetables | Grey mould rot | *Botrytis cinerea* Pers. ex. Fr. |
| Papaya, mango | Anthracnose | *Colletotrichum gloeosporiodes* (Penz.) Sacc. |
| Peach, cherry | Brown rot | *Monilinia fructicola* (Wint.) Hone-; (= *Sclerotinia fructicola* [Wint.] Rehm) |
| Peach, cherry, strawberry | Rhizopus rot | *Rhizopus stolonifer* Ehr. ex Fr. |
| Pineapple | Black rot | *Ceratocystis paradoxa* (Dade) Moreau (= *Thielaviopsis paradoxa* [de Seynes] Hohn.) |
| Potato, leafy vegetables and other species | Bacterial soft rot | *Erwinia carotovora* (Jones) Holland . |
| | Dry rot | *Fusarium* spp. |
| Sweet potato | Black rot | *Ceratocystis fimbriata* Ellis and Halst. (=*Endoconidiophora fimbriata* [Ell. and Halst.] Davidson) |
| Leafy vegetables, carrot | Watery soft rot | *Sclerotinia sclerotiorum* (Lib). de Bary |

SOURCES J.W. Eckert (1977) Control of postharvest diseases, in M.R. Siegel and H.D. Sisler (eds) *Antifungal compounds*, vol. 1, Marcel Dekker, New York, pp. 269–352; J.W. Eckert and G.E. Brown (1986) Postharvest citrus diseases and their control, in W.F. Wardowski, S.C. Nagy and W. Grierson (eds) *Fresh citrus fruits*, AVI, Westport, CT, pp. 315–60.

similar because they are effective against the same range of fungi. The benzimidazoles inhibit spore germination, interfere with mycelial growth and affect conidia formation even if the fruit are treated too late to control infection. It is thought that the benzimidazoles interfere with polymerisation of the protein tubulin. This very specific action is responsible for the development of resistant strains of *Penicillium*.

The benzimidazoles are not readily soluble in water and are marketed as wettable powder formulations. Citrus fruit are generally flood-irrigated with a suspension of 0.05–0.1 per cent of the selected benzimidazole for 30 seconds to ensure thorough contact of the fungicide with sites of injury (Figure 9.2 in the colour plate section). Fungicides may also be incorporated in wax emulsions (Figure 9.3 in the colour plate section). The benzimidazoles have substantial residual activity. For example, benomyl is able to provide excellent wastage control in lemons stored for 6 months at 13°C. Because of the ease with which these compounds may be handled, packing houses rapidly adopted the benzimidazoles, but the development of resistant strains of *Penicillium* has caused many packing houses to look for alternative treatments.

Some of these newer fungicides are described below.

*Sec-butylamine (2-AB)*: This is effective against benzimidazole-resistant strains, but has now been withdrawn from use because of lack of re-registration and subsequent production. It was effective as a water treatment or fumigant. Resistant strains of mould have also been reported to this compound, with some isolates even being resistant to the benzimidazoles as well.

*Guazatine*: This fungicide is extensively used in Australia and New Zealand and has the ability to control citrus *Penicillium* moulds and sour rot. It has been especially valuable in controlling moulds that have been resistant to the benzimidazoles. Multiple resistance has, however, been reported to this compound and the benzimidazoles, but to date this has not emerged as a serious commercial problem as the strain does not compete well with the 'wild' type.

*Triazoles*: The fungicides imazalil and prochloraz are members of this group and are very effective against a wide range of fungi, particularly some of those that are resistant to the benzimidazoles. These compounds act by inhibiting the demethylation process during ergosterol biosynthesis within the fungus. The fungicides are very effective against a wide range of moulds and also prevent sporulation of the fungus if decay occurs. They can also act as a fumigant with limited efficacy. Like the previous compounds mentioned, resistance has now been reported in Californian lemon storage rooms. Prochloraz is also registered for use on mango for the control of anthracnose.

## Development of chemical treatments to control wastage in other fruits

The storage and transportation of other fruits benefit from suitable selection and application of chemical fungicides. For example, black-end, finger-stalk or crown rot of bananas and brown rot in peaches are effectively controlled by the benzimidazoles; grey mould in table grapes by the slow evolution of sulphur dioxide from an in-package generator or by room fumigation with sulphur dioxide; and *Rhizopus* rot in stone fruits by dicloran. In contrast, control of *Geotrichum*, *Alternaria* and the soft rots is still unsatisfactory, although SOPP does give some control of *Geotrichum* when applied 4–24 hours after infection. Furthermore, much more needs to be learned about the mechanisms that lead to the appearance of strains resistant to some of the more potent, but quite specific in action, fungicides.

In summary, the ideal postharvest fungicide should be water soluble, have broad spectrum activity, not be phytotoxic, be safe to use (i.e. leave no residues that are toxic to consumers), not affect palatability, retain activity over a long period, leave no visible residues and be cheap (i.e. the compound should be inexpensive or effective at low concentration). None of the present fungicides meet all of these requirements.

## Non-chemical methods of decay control

There is close regulation of the use of fungicides, which must pass thorough checking and approval by the appropriate government authorities in each country to ensure they are not toxic at the concentrations used, to set maximum residue limits (MRL) for approved fungicides and to empower government inspection services to monitor that the products are only being used for the specific approved purpose. Despite this protocol, there is growing concern and apprehension by the public about the use of synthetic 'chemicals', and pressure is building for the use by the horticultural and agricultural industries of alternative 'non-chemical' means of disease control. Desirable alternative treatments are perceived to be physical treatments or biological control as they are considered more natural and avoid residue problems arising when chemical fungicides are used.

### Heat treatment

The concept of killing pathogenic fungal spores by heat treatment is not new. In the early 1930s, fruit were passed through hot dips for up to

4 minutes at 49°C to kill mould spores on the citrus fruit rind. Heat treatment was generally used in conjunction with older-style fungicides such as sodium bicarbonate (baking soda) to enhance its activity. Later studies in 1980 by the NSW Department of Agriculture in Australia showed that 99 per cent of green mould spores would be killed after 1 hour exposure to hot water at 50°C, thus preventing the build-up of a viable spore population in these washing tanks. What is new is the incentive to use non-chemical means of mould control. Heat treatments, however, not only affect the pathogen but can have beneficial effects on the fruit. Research in Israel has shown that if citrus fruit are held at 35°C in a humid environment (95–99 per cent RH), mould infection does not occur and decay is prevented. This is due to enhanced formation of lignin and related compounds that prevent invasion by germinating mould spores. The use of heat needs to be carefully controlled as problems can be encountered with excess heat application leading to enhanced ageing of the fruit and hence a reduction in quality.

## Biological control methods in fruit

While there are potentially many biological control agents that may be useful in controlling postharvest wastage — including fungi, yeasts and bacteria — identification and commercial evaluation is at a relatively infant state.

The ability of natural antibiotic substances to control mould growth is becoming better known. Control of the fungal disease *Rhizopus stolonifer*, the cause of transit rot in peaches, has been achieved with very high concentrations of the bacterium *Enterobacter cloacae* ($10^{12}$ bacteria/mL). The bacterium *Bacillus subtilis* has been shown to control brown rot of peaches when fruit were sprayed with a suspension of $10^9$ live bacteria per mL. *B. subtilis* has also been found as an active antagonist against citrus green mould, sour rot and *Alternaria* centre rot. The controlling agents are assumed to be antifungal substances produced by the bacterium that prevent mould development. Brown rot of peaches has also been reduced by two as-yet unidentified antibiotic substances isolated from the growth media of the fungus *Penicillium frequentans*.

Another method of biological control is where disease can be controlled by organisms that feed on the pathogen. This mechanism has been reported as a method of controlling the root-rotting fungus *Sclerotinia* by parasitism of the resting stage, the sclerotia, by the fungus *Coniothyrium*. Studies in Australia also found a fungus *Paecilomyces* engulfed a culture of the fungus *Geotrichum candidum* that was growing on an agar plate (Figure 9.4 in the colour plate section). Unfortunately this organism did not control sour rot when its spores were applied to inoculated fruit.

Competition between organisms is a promising area of biological control, with the growth of a non-pathogenic organisms preventing the growth of a pathogenic organism. The yeast *Debaryomyces* (now called *Pichia*) appears to reduce green mould development by competing for space and nutrients in an injury site on the fruit rind of citrus and thus inhibiting mould development. It does not produce an antibiotic, which has advantages as such compounds may have toxic effects on consumers.

While antifungal substances or the organisms themselves may occur naturally, they will have to be tested for human toxicity and cancer risks in the same manner that currently used fungicides are screened. The biological control methods may have additional problems with allergic responses in humans.

### *Future direction for research*

The major research thrust around the world is towards alternative, non-chemical means of disease control. In addition to improvements to physical methods and control with biological agents, considerable interest is being generated into naturally occurring chemical compounds that have anti-fungal properties. Many of these compounds, such as ethanol and hexanal, are volatile at storage temperatures and present opportunities as vapours released into the atmosphere around produce during storage or transport. These volatile compounds are produced naturally at quite low levels but, to be effective, commercial treatments will need to have their concentration elevated by enhanced synthesis or by exogenous application. The regulatory process for naturally occurring compounds that are used at non-natural, elevated concentrations is still to be determined.

However, like most disease control programs, decay prevention should not just be based on one aspect of control but on a whole integrated program of good hygiene and careful handling. These latter two procedures, possible by themselves, are the most effective, non-chemical, safe way of reducing decay development in fruit.

# FURTHER READING
·······

Barkai-Golan, R. and D.J. Phillips (1991) Postharvest heat treatment of fresh fruits and vegetables for decay control, *Plant Dis.* 75: 1085–89.

Beattie, B.B., W.B. McGlasson and N.L. Wade (1989) Postharvest diseases of horticultural produce, vol. 1, *Temperate fruit*, CSIRO, East Melbourne.

Coates, L., T. Cooke, D. Persley, B. Beattie, N. Wade and R. Ridgway (1995) Postharvest diseases of horticultural produce, vol. 2, *Tropical fruit*, Department of Primary Industries, Queensland, Brisbane.

Coates, L.M. and G.I. Johnson (1993) Effective disease control in heat disinfested fruit, *Postharv. News Inform.* 4: 35N–40N.

Dekker, J. and S.G. Georgiopoulos (eds) (1982) *Fungicide resistance in crop protection*, Pudoc, Wageningen.

Eckert, J.W. and G.E. Brown (1986) Postharvest citrus diseases and their control, in W.F. Wardowski, S.C. Nagy, W. Grierson (eds), *Fresh citrus fruits*, AVI, Westport, CT, pp. 315–60.

Erwin, D.C. (1973) Systemic fungicides: Disease control, translocation, and mode of action, *Annu. Rev. Phytopath.* 11: 389–422.

Lieberman, M. (1983) *Post-harvest physiology and crop preservation*, Plenum, New York.

Reuther, W., E.C. Calavan and G.E. Carman (eds) (1978) Citrus industry, vol. 4, *Diseases and injuries, viruses; registration, certification, indexing; regulatory measures, vertebrate pests; biological control of insects; nematodes*, University of California, Berkeley, CA.

Snowdon, A.L. (1990) A colour atlas of post-harvest diseases and disorders of fruits and vegetables, vol. 1, *General introduction and fruits*, Wolfe, London.

— — (1991) A colour atlas of post-harvest diseases and disorders of fruits and vegetables, vol. 2, *Vegetables*, Wolfe, London.

Wilson, C.L. M.E. Wisniewski (1994) *Biological control of postharvest diseases*, CRC Press, Boca Raton, FL.

# 10

# EVALUATION AND MANAGEMENT OF QUALITY

The term 'quality' defies complete and objective definition. For each consumer of horticultural produce, quality is a highly subjective judgement related to learned criteria. For example, orange fruit can be afflicted by a mite that causes the skin to go bronze. Some individuals 'in the know' register the bronzed appearance (quality criterion) as a promise of a very sweet orange, while most might register it as the appearance of a fruit to be avoided.

With respect to fruit, vegetables and ornamentals, quality criteria will, of course, vary between commodities. Furthermore, for any particular commodity, the definition and associated criteria will also depend on the perspective of the recipient in the handling chain (e.g. grower versus marketer versus consumer). For example: what is a good quality pear?

- To the producer, a good quality pear is one that secures a maximum price in the market at a particular time of the season. The producer must choose between either a high quality but (usually) low yield product requiring intensive care or a poorer quality product with a higher yield through less intensive care. For example, a higher total return may be gained for pears picked early despite the fact that the quality may not be ideal for the consumer and the yield on a weight basis will be generally lower than for a later picked crop.
- To the shipper, a good quality pear is a hard green pear that is capable of being transferred from orchard to market without bruising or ripening — the harder the better!
- To the canner, good quality is a ripe but firm pear. The consumer requires a canned pear to be soft, but the pear needs to be firm enough to retain its shape undamaged during processing and subsequent market handling.
- To the consumer of the fresh fruit, a good quality pear is a soft, ripe product. It must melt in the mouth and be juicy. Skin colour is also often important, whereas to the canner it is immaterial as the skin is removed prior to processing.

**159**

Quality may therefore be defined in terms of end use or 'fitness for purpose'. In this context, produce quality requirements commonly refer to market, storage, transport, eating and/or processing quality. The marketing of fresh fruit and vegetables and of ornamentals is aimed eventually at appealing to the consumer for whom tradition (i.e. learned criteria) plays a major role in determining acceptability. The purchasing habits of people are typically conservative so that an inducement is often required to get people to try something new, such as a pear with a red skin.

Because of markedly different quality considerations between ornamentals and produce (fruit and vegetables) grown for eating, postproduction quality evaluation for ornamentals is considered under a separate heading later in this chapter.

# QUALITY CRITERIA

The criteria of quality can be divided into external and internal factors. In terms of selling produce, external criteria might be considered of paramount importance. Accordingly, measures are often taken to try to 'improve' external quality. Examples of such measures include waxing of apples, degreening of oranges, reddish lighting to enhance the colouration of red apples, and orange coloured mesh bags to reinforce the colouration of oranges. It is debatable whether such practices are ethical or not. However, if the consumer is disappointed by poor internal quality, eventually a reduction in the number of repeat purchases can be expected. For instance, consumers are often dissappointed by poor organoleptic properties of early season fruit (e.g. immature nectarines that fail to ripen properly) or out of season fruit (e.g. mealy apples that have been overstored). Important quality criteria for consumers are:

♦ appearance, including size, colour and shape;
♦ condition and absence of defects;
♦ mouthfeel or texture;
♦ flavour; and
♦ nutritional value.

## *Appearance*

People 'buy with their eyes' and learn from experience to associate desirable qualities with a certain external appearance. A rapid visual assessment can be made on the basis of size, shape, colour, condition (such as freshness) and/or the presence of defects or blemishes.

Size is an important criterion of quality that can be easily measured by circumference, diameter, length, width, weight or volume. Many fruit are graded according to size, often by diameter measurement, with similar sizes of fruit being packed together to give a uniform pack that assists marketing and retail sales (Figure 10.1). For example, certain size standards are adhered to for export apples based on fruit diameter or associated count number per package. These standards are specific to a particular package, and also depend on their export destination. The maximum size limits for apples ensure that large apples, which are highly susceptible to postharvest physiological disorders such as storage breakdown, are not exported. Particular packing arrays and individual wrapping of fruit, especially the top layer, may also be mandatory or recommended. The packing array will determine the number of fruit per package for some commodities. An example of a commodity graded by length and diameter is the carrot. Weight is the standard determinant for many other commodities.

Shape is a criterion that often distinguishes particular cultivars of fruit. Characteristic shapes are usually demanded by the consumer, who will often reject a commodity that lacks the characteristic shape. For example, attempts to market a straight banana were unsuccessful, apparently because this shape was considered abnormal. Shape is especially

**Figure 10.1** Modern machinery for sizing, colour sorting and packing fruit. In this example a colour imaging system is used to size and colour grade the fruit. (Courtesy of Colour Vision Systems P/L, Victoria, Australia.)

important in apple cultivars, a premium price being obtained for fruit with a well developed shape that is characteristic of the particular cultivar. Mis-shapen fruit and vegetables are poorly accepted. Any deviation in shape usually brings a lower price. Shape is often a problem in breeding programs. Although a superior eating or storing product may be obtained by breeding a new cultivar, if its shape is unusual it will be less readily accepted in the market and will require extensive re-education of the consumer through advertising programs.

One of the distinguishing features of fruit and vegetables is that they are the only major group of natural foods with a variety of bright colours. They are therefore often used merely to brighten up the presentation of foods. Parsley contains relatively high levels of ascorbic acid, carotene, thiamin, riboflavin, iron and calcium compared with fruit and other vegetables, yet is used almost exclusively in Western society to add colour and flavour to meat and fish dishes. A bright red apple is most valued in some markets, yet the colour of the skin adds nothing to its eating quality or to its nutritional value. Colour changes in ripening fruit have been associated by the consumer with the conversion of starch to sugar (i.e. sweetening) and the development of other desirable attributes, so that the correct skin colour is often all that is required for a decision to purchase the commodity. Such subjective assessments may be misleading. For example, if fruit such as the banana are ripened at higher than optimum temperatures, full loss of green colour does not occur and consumers, particularly those from temperate climates, would show considerable buyer resistance even though the flesh is adequately ripened. Standardised colour charts are used in the visual assessment of ripeness in many fruit, such as tomatoes, pears, apples and bananas (Figure 10.2 in the colour plate section).

### Condition and defects

The condition of a commodity is a quality attribute referring usually to freshness, stage of senescence or ripeness, the extent of mechanical damage and pest or disease incidence. Wilted leafy vegetables obviously lack condition and are generally unacceptable to the consumer. Similarly, fruit that are shrivelled, due to excessive loss of moisture, also lack condition. The loss of condition is most marked towards the end of the shelf-life of a commodity in the retail outlet. Prevention of such loss of condition can be achieved by improved storage conditions, which usually equates with cool storage but may be as simple as shading the produce, especially leafy vegetables, from direct sunlight.

Skin blemishes, such as bruises, scratch marks and cuts, detract from appearance and in most markets detract from price, even when the blemishes reduce neither keeping quality nor eating quality. Nonetheless, although a premium price may be obtained for produce that is free from blemishes, there will still often be a market for lower grade produce (e.g. through roadside stalls). Consumers in developing countries, for a variety of reasons, particularly economic, tend to be less discerning than those in more developed countries. Consequently a greater proportion of a crop that reaches the market in developing countries is likely to be sold and consumed.

Not all consumer correlations are necessarily based on valid scientific evidence. Some commodities such as bananas frequently develop latent infections of the skin during ripening as a consequence of the weakening of the skin structure. Thus, the ripe product is often flecked (i.e. yellow with black spots). If this mould growth is controlled, as is now possible, the product may then be regarded by some consumers with suspicion ('it doesn't look right!') even though eating quality of the flesh is unaffected. Normal appearance is extremely important in the market place. Acceptable appearance, however, may differ between countries and between different regions within a country. In Japan, great importance is attached to the unblemished appearance of the fruit. For example, a perfectly even yellow colour is required in banana peel. Similarly, only 'netted' melons without a 'ground spot' are in demand in Japanese markets.

Thus, appearance is a major determinant of quality, especially because it is often the only criterion available to the buyer of the commodity. On-the-spot taste testing is rarely practised or encouraged at the retail level, although it is more general at the commercial markets from which the retailers purchase their commodities.

## *Mouthfeel*

Mouthfeel, including texture, is the overall assessment of the feeling the food gives in the mouth. It is a combination of sensations derived from the lips, tongue, walls of the mouth, teeth and even the ears. Each of these areas is sensitive to small pressure differences and responds to different attributes of the produce. Lips sense the type of surface being presented; for instance, they can distinguish between hairy (pubescent) and smooth (glaborous) surfaces. Teeth are involved in determining rigidity of structure. They are sensitive to the amount of pressure required to cleave the food and to the manner in which the food gives way under the applied force. The tongue and walls of the mouth are sensitive to the types of particles generated following cleavage by the teeth (e.g. whether they are soft and mushy or discrete lumps). The extent of expressed juice is also assessed. Ears sense the sounds of the food being chewed, and intimately complement mouthfeel. Sound is vitally important in produce such as celery, apples and lettuce, where crispness is a critical attribute. The cumulative effect of these responses creates an overall impression of the mouthfeel of the produce.

## Flavour

Flavour is comprised of taste and aroma. Taste is due to sensations felt on the tongue. The four main taste sensations are sweet, salt, acid (sour) and bitter. Each sensation is largely perceived at a specific area of the tongue. All food tastes elicit a response on one or more areas. The taste of fruit and vegetables is usually a blend or balance of sweet and sour, often with overtones of bitterness due to tannins. They are not naturally salty. Aroma is due to stimulation of the olfactory senses by volatile organic compounds, some of which have been outlined previously in Chapter 2.

## Nutritional value

Nutrition is probably the least important consideration in determining whether a consumer purchases a commodity, since most essential nutrients can neither be seen nor tasted. Fruit and vegetables are the sole source of vitamin C in the diet of many people. In days gone by, British seamen carried limes on board their ships and consumed them as a precaution against the disease scurvy. However, nowadays, few people would decide to buy a particular piece of fruit because it had more vitamin C than another type of fruit. Even so, the nutritional properities of crops such as avocados have been sucessfully promoted as containing no cholesterol, while orange juice is widely perceived as being high in vitamin C.

Improved nutritional value should be the aim of anyone connected with any aspect of the fruit and vegetable industry as it is a means of upgrading the health of the community without changing their food habits. However, grower and consumer acceptance of varieties with superior nutritional value has been marginal. Some type of government subsidy or coercion is imperative if nutritional gains are to be made within a reasonable time. This was well demonstrated with corn (maize). Breeding programs developed corn varieties with high levels of the amino acids lysine and tryptophan, thus overcoming a serious nutritional deficiency in normal corn varieties. However, early acceptance, and therefore the benefit, was limited. The new varieties looked different and had some different cultural requirements so that both farmers and consumers saw no benefit in the new corn.

Population groups with impoverished diets tend to be poorly educated. Therefore it is difficult to convince them of the benefits to be gained from a change in diet. Nonetheless, progress is being made in increasing awareness of the importance of a nutritious and balanced diet, especially among people in 'developing countries' through the efforts of international development organisations.

# POSTHARVEST FACTORS INFLUENCING QUALITY

By no means do all changes in harvested produce equate to loss of quality. Many of the physicochemical changes that occur after harvest are essential for the desired degree of eating quality to be attained. In general, climacteric fruit such as bananas, mangoes and tomatoes are picked at the mature-green stage of development and then allowed to ripen off the plant to optimum eating quality. Harvest at the mature-green stage is almost mandatory for avocado fruit, which will not ripen while attached to the tree. It is not until very late in the season that avocado fruit will detach naturally, with virtually concomitant ripening. Nonetheless, the principal concern with many other harvested horticultural produce — such as non-climacteric fruit, immature and leafy vegetables and cut flowers and foliage — is preventing loss of the existing quality. This deterioration in quality can be caused by a variety of stresses that may be grouped into four general, but often inter-related, categories: metabolic stress, transpiration, mechanical injury stress and microbial damage.

Metabolic stress involves either 'normal' or 'abnormal' metabolism that lead to senescence or the development of physiological disorders, respectively. Respiration-induced cabohydrate shortage in cut flowers and heat treatment-induced impairment of starch breakdown in mangoes are, in turn, examples of normal and abnormal metabolism. While the loss from physiological disorders is often spectacular (e.g. extensive skin browning in bananas as a result of severe chilling injury), such disorders are generally far less common problems than normal senescence *per se*. Furthermore, certain aspects of general senescence can be both as rapid and dramatic as manifestations of physiological disorders, such as chlorophyll degradation in harvested broccoli.

Transpiration and subsequent water loss can also result in rapid loss of quality (as well as direct loss in saleable weight and therefore absolute monetary value). Severe wilting of leafy vegetables and cut flowers and foliage can be induced by storage for a matter of hours under hot, dry conditions. Water loss mainly affects appearance, through wilting and shrivelling, and texture, such as loss of crispness of lettuce. However, water loss can also affect nutritional quality. For instance, vitamin C levels fall rapidly in water-stressed leafy vegetables.

Mechanical injury causes loss of visual quality characterised by unsightly abrasions, bruises, cuts and tears. Such injuries lead to an increase in the general metabolic rate (wound response) as the produce tries to seal off the damaged tissues. Furthermore, transpiration increases because natural barriers against the loss of water (e.g. the cuticle) have been damaged. Such injury is not always the result of an unintentional action but may be a side-effect of a deliberate postharvest treatment, such as damage caused to the cuticle when peaches are brushed to remove the fuzz (trichomes) from the skin.

Microorganisms can often be considered a 'secondary stress', since their proliferation is generally facilitated by mechanical injury, transpiration and/or metabolic changes such as senescence and physiological disorders. This relationship is particularly true for the weak saprophytic organisms that cause postharvest rots, such as *Alternaria alternata*. Anthracnose disease of avocado fruit, which is caused by *Colletotrichum gloeosporioides*, becomes evident during ripening when levels of natural antifungal compounds, such as dienes, fall below fungistatic concentrations. Consequently, many microbial problems can be minimised or eliminated by careful and proper postharvest handling practices. Postharvest diseases are mainly caused by fungi, although some bacteria and yeasts are also pathogenic. Where favourable environmental conditions of temperature, pH and water status for their growth prevail, the growth of pathogens can be extremely rapid and results in extensive losses (see Chapter 9).

Some of the major handling factors that contribute to loss of quality of harvested produce are discussed below.

### Harvesting

Mechanical damage during harvesting and associated handling operations can result in defects on the produce and permit invasion by disease-causing microorganisms. The inclusion of dirt from the field can aggravate this situation. Produce can overheat and rapidly deteriorate during temporary field storage. Failure to sort and discard immature, overripe, undersized, misshapen, blemished or otherwise damaged produce creates problems in the subsequent handling and marketing of the produce.

### Transport and handling

Rough handling and transport over bumpy roads damages produce by mechanical action. At high temperatures produce will become overheated, especially if there is inadequate shading, ventilation and/or cooling. Transport on open trucks can result in sun-scorch of the exposed produce. Severe water loss, especially from leafy vegetables, can also occur under these conditions. Inappropriate packaging (e.g. overfilling, underfilling) may result in physical damage of produce due to bruising or to abrasion as the commodity moves about during transport. Temperature variations can lead to condensation, which in turn may encourage decay and weaken packages.

## Storage

Delays in placing produce in cool storage after harvest often result in rapid deterioration in quality. Poor control of storage conditions, storage for too long and inappropriate storage conditions for a particular commodity will also result in a poor quality product. With mixed storage of different commodities, ethylene produced by one product (e.g. ripening fruit) can promote rapid senescence of another product (e.g. leafy vegetables). Storage at temperatures that are too low may induce physiological disorders or chilling injury. High temperature and high humidity can encourage both superficial and internal mould growth and stimulate activity of infesting insects.

## Marketing

A serious reduction in quality can occur in produce displayed for lengthy periods in retail outlets because of poor organisation of marketing. Major causes of quality reduction during marketing include ongoing growth (e.g. opening of cut flowers), water loss leading to wilting, undesirable ripening (e.g. softening of apples) and senescence (e.g. loss of green colour and yellowing of leafy vegetables) under poor management of temperature and RH, mechanical damage associated with rough handling by staff and customers, and associated disease development. An interesting marketing problem is the greening of potatoes, which is caused by the practice of displaying cleaned potatoes in relatively bright light in supermarkets. Greening is strongly associated in the mind of the consumer with the accumulation of solanine, a toxic glycoalkaloid.

## Treatment residues

Residues of pesticides and other chemicals are another important factor impinging upon postharvest quality. Chemicals such as insecticides and herbicides are often applied preharvest. Fungicides may be used both preharvest and/or postharvest to prevent rotting (Chapter 9). Fumigants such as methyl bromide may be used for insect disinfestation, especially in export trade or disease control (e.g. slow release sulphur dioxide pads). All of these types of chemicals can leave residues in the commodity that, although usually not detectable by the consumer, is certainly considered undesirable by the consumer and must be considered in conjunction with possible health risks to the community. The use of electron beam and gamma irradiation to inhibit mould growth or sprouting may reduce chemical residues but can still create an image problem with consumers.

# DETERMINATION OF MATURITY

Maturity is an integral component of quality, especially in the context of commecial maturity. There is a clear distinction between 'physiological' and 'commercial or horticultural' maturity. The former is a particular stage in the development of a plant or plant organ, and the latter is concerned with the timing of the harvest to meet particular market requirements. At optimum commercial maturity, produce should be either at optimum consumer quality (e.g. ripe in the case of non-climacteric fruit such as oranges) or able to achieve optimum consumer quality (e.g. at an advanced bud stage in the case of potted chrysanthemum plants intended for the Mothers' Day market).

Physiological maturity refers to the point in the development of an organ (e.g. fruit, leaf) or organism (e.g. ornamental plant) when maximum growth has been achieved and the organ or organism has matured to the extent that the next development stage can be completed. In the case of fruit, ripening can be considered the next development stage, preceding the senescence stage. Clear distinctions between the stages of development — namely growth, maturation, ripening and senescence — of a plant organ or organism are generally not easy (see Chapter 3) because transitions between the various development stages are often slow and indistinct. Nevertheless, in fruits in particular, measurement of physiological (e.g. respiration and ethylene production) and/or biochemical characteristics (e.g. sugar/acid ratios) can give reliable estimates of the degree of maturity for specific commodities (see Chapter 11).

Commercial maturity is the characteristic state of a plant organ required by a market. Commercial maturity commonly bears little relation to physiological maturity, and may occur at any stage during development, maturation, ripening or senescence. Examples of commercial maturity include bean sprouts (during development), cucumbers (during maturation) and strawberries (during ripening/senescence). The terms 'immaturity', 'optimum maturity' and 'overmaturity' can be related to these market requirements. Thus, there must be understanding of each in physiological terms, particularly where storage life and quality when ripe are concerned. Some examples of commercial maturity in relation to physiological age are shown in Figure 10.3.

Ripening, as applied to fruit, is the process by which a fruit attains its maximum desired eating quality. The ripening process presumably evolved as a means of attracting fruit-eating animals in order to ensure dispersal of the seeds within fruit. Before the ripe stage, a fruit is said to be under-ripe, and later it becomes over-ripe. The three ripeness conditions cannot be clearly defined physiologically because they are subjective judgements, and thus will vary among consumers. The under-ripe condition overlaps the mature stage of development, while the over-ripe condition overlaps the senescent stage of development. Consider the banana, which some relish as soon as it yellows but which others eat only when brown flecks appear on the skin. The potential ripe quality is determined by many factors, of which the stage at which the fruit was harvested is vitally important. Fruit such as nectarines will not ripen to attain desirable or optimum organoleptic properties when

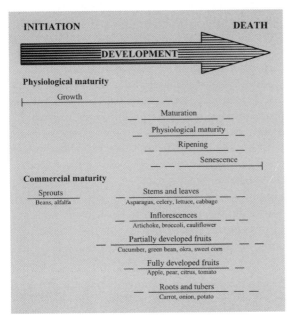

**Figure 10.3**
Commercial maturity in relation to developmental stages of the plant.
SOURCE Adapted from A.E. Watada, R.C. Herner, A.A. Kader, R.J. Romani and G.L. Staby (1984). Terminology for the description of developmental stages of commercial crops, *HortScience* 19: 20–21.

picked immature while bananas can be satifactorily harvested over a wide range of maturity.

# DETERMINATION OF COMMERCIAL MATURITY

Commercial maturity indices generally involve some expression of the stage of development (growth, maturation or ripening) and usually require the determination of some characteristic known to change as the plant material matures. Indices may involve decisions about levels of market and consumer acceptability, and generally necessitate objective measurements, subjective judgements or both. Objective and subjective assessments may be destructive or non-destructive in nature.

The determination of the time of harvest for peas for processing by canning or freezing provides an interesting example of the need for close monitoring of commercial maturity. Destructive objective measurements of maturity in shelled peas, beans and broad beans can be obtained using any of several instruments specially designed for this purpose. They operate on the principle of the force required to shear the seed, and include the use of the tenderometer (used mainly in the USA and UK) and the maturometer (used mainly in Australia). The end result of early research and development aimed at overcoming problems in the determination of optimum maturity of shelled vegetables for processing has been the provision for the consumer of a generally uniform, high quality canned and frozen product. Such objective standards cannot always be set for maturity determination in many commodities. In the final analysis, judging the time of harvest is often based on the growers' experience with their own crops in terms of calendar date and various subjective judgements in relation to market requirements. Cut roses are a good example. Those destined for local markets can be harvested at a more advanced stage, with the outermost petal lifting from the bud. In contrast, those destined for storage and transport to distant markets are harvested at a considerably 'tighter' bud stage.

Many criteria for judging maturity (Table 10.1) have been used or suggested based on a variety of characteristics including:

◆ time from flowering or planting (calendar date);
◆ accumulated heat units;
◆ size and shape;
◆ skin or flesh colour;
◆ light transmission or reflection;

- flesh firmness;
- electrical conductance or resistance;
- chemical composition (e.g. starch, sugar, acid);
- respiratory behaviour and ethylene production; and
- time to ripen.

To be practical, maturity tests should ideally be simple, rapid and readily applied in the field. If they are non-destructive, so much the better. A description of these methods follows.

## Calendar date

For perennial fruit crops grown in seasonal climates that are more or less uniform from year to year, calendar-based harvest dates can be a useful guide to commercial maturity. This approach relies on a reproducible date for the time of flowering and a relatively constant growth period from flowering through to maturity. Time of flowering largely depends on temperature, and the variation in the number of days from flowering

---

### TABLE 10.1 ESTABLISHED METHODS FOR EVALUATION OF MATURITY IN HORTICULTURAL PRODUCE

| MATURITY INDEX | PRODUCE |
| --- | --- |
| abscission | rockmelons |
| accumulated heat units | peas |
| astringency | persimmon |
| chronological time | apples |
| colour | tomatoes |
| stage of development (e.g. density) | lettuce |
| dry matter content | mangoes |
| firmness | peas |
| internal ethylene | apples |
| juice content | oranges |
| oil content | avocados |
| shape | bananas |
| size | zucchini |
| starch–iodine staining | apples |
| sugar, acid and sugar/acid by refractometer and titration | oranges |
| waxiness and gloss | grapes |

to harvest can be calculated for some commodities by use of the degree-day concept (see below). Calendar-based harvest dates are usually derived from growers' experience with crops in a particular environment.

## Heat units

An objective measure of the time required for the development of the fruit to maturity after flowering can be made by measuring the 'degree days' or 'heat units' in a particular environment. It has been found that a characteristic number of heat units or degree days is required to mature a crop. Maturity will be advanced under unusually warm conditions or delayed under cooler conditions. The number of degree days to maturity is determined over a period of several years by obtaining the algebraic sum of the differences between the daily mean temperatures and a fixed base temperature (commonly the minimum temperature at which growth occurs). For example:

$$20 - 15°C \text{ (minimum) for 1 day} + 25 - 15°C \text{ for 2 days} = 15 \text{ degree days}$$

The average or characteristic number of degree days is then used to forecast the probable date of maturity for the current year (e.g. 1000 deg.days) and, as maturity approaches, it can be checked by other means. This heat unit approach is extremely helpful in planning planting, harvesting and factory management programs for annual processing crops such as corn, peas and tomatoes. Similarly, this approach is useful in programming cut flower or flowering pot plant production (e.g. lilies for the Easter market or poinsettias for the Christmas market).

## Shape and size

Fruit shape may be used in some instances to decide maturity. For example, some cultivars of bananas become less angular in cross-section as development and maturation progress. The fullness of the 'cheeks' adjacent to the pedicel may be used as a guide to the maturity of mangoes and some stone fruits. Size is generally of limited value as a maturity index in fruit, although it is widely used for many vegetables, especially those marketed early in their development such as zucchini. Size, in combination with shape, is also an important determinant of commercial maturity in ornamentals such as flowering and foliage pot plants. With such produce, size is often specified as a quality standard. Large size generally indicates commercial overmaturity and undersized produce indicates an immature condition. This assumption, however, is, not always a reliable guide for all produce.

Shape and size give rise to the volume parameter, which when expresssed as a function of weight (mass) gives density (e.g. $g/cm^3$). This ratio is a useful determinant of maturity in produce such as potatoes. It can also be used to sort defects (e.g. disorders that result in internal cavities).

## *Colour*

The loss of green colour — often referred to as the 'ground colour' (i.e. the background colour) — of many fruits is a valuable guide to maturity. There is initially a gradual loss of intensity of colour from deep green to lighter green and, with many commodites, complete loss of green with the development of yellow, red or purple pigments (e.g. tomatoes). Ground colour, as judged against prepared colour charts, is a useful index of maturity for apples, pears and stone fruits, but is not entirely reliable as it is influenced by factors other than maturity (e.g. nitrogen uptake during growth).

For some fruits, development of blush colour (i.e. additional colour superimposed on the ground colour) as they mature on the tree can be a useful indicator of maturity. Examples are the red or red-streaked apple cultivars and red blush on some cultivars of peach. Such colour development usually depends on exposure to sunlight. For other fruits, such as the characteristically green apple cultivar Granny Smith, over-exposure to sunlight leads to an undesirable yellow sunburn lesion, which diminishes market value. For certain tropical fruits such as papaya (pawpaw), skin colour is a reliable guide to commercial maturity. The appearance of a trace of yellow at the apical (distal) end of papaya fruit may be used to determine the time of harvest. However, such fruit would benefit from a longer period on the tree, until about one-third of the fruit is yellow. After this time, papaya fruit ripening is quicker and eating quality is better, but the shelf-life will be shorter. Thus, the grower must balance the benefits of improved eating quality against greater risk of losses associated with over-ripening during marketing. For table market tomatoes, both the loss of green colour (chlorophyll) and the development of red colour (lycopene) are integrated in pictorial charts by which ripeness is judged (Figure 10.4 in the colour plate section). In contrast, external colour can be useless as a guide to maturity for oranges grown in tropical areas, which remain green or may re-green, and for pineapples, the flesh of which may be mature even though the shell is green.

Objective measurement of colour is possible using a variety of light reflection or transmission spectrophotometers. The Hunter and Minolta colour/colour difference meters, which measure surface colour, are widely used in research work. Colour video image analysers can also be used to sort horticulture produce on the basis of both colour and

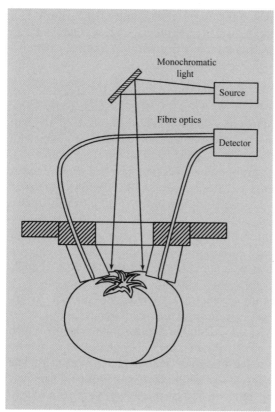

**Figure 10.5**
Diagram of the geometry used to make transmission measurements of fruit. The incident light that enters the product directly under the source is scattered and absorbed by the product. Some of the scattered light transmitted by the product is collected by the shielded detectors placed directly on the surface of the product. Transmission spectra are generated by varying the wavelength of the incident light.
SOURCE Adapted from G.S. Birth, G.G. Dull, J.B. Magee, H.T. Chan and C.G. Cavaletto (1984) An optical method for estimating papaya maturity, *J. Am. Soc. Hort. Sci.* 109: 62–69.

dimensions (Figure 10.1). Electronic colour vision sorting is now common for apples, citrus and tomatoes. As the capital cost of sophisticated instrumentation is high, its use is largely restricted to packing houses with high throughput of produce. Nonetheless, as labour costs increase, broader (e.g. field) application of such instrumentation can be anticipated. Changes in internal colour and other physicochemical characteristics can be gauged using broad (e.g. white) or narrow (e.g. infrared) band light applied at the surface, such as with transmission spectrophotometers (Figure 10.5).

## Flesh firmness

As fruit mature and ripen they soften, largely because the pectins comprising the middle lamella of cell walls dissolve. This softening can be estimated subjectively by finger or thumb pressure. However, more objective measurement, yielding a numerical expression of flesh firmness, is possible with a fruit pressure tester (penetrometer). These testers measure the pressure at which flesh yields to the penetration of a standard diameter plunger inserted to a standard depth. Commonly used penetrometers are the Magness–Taylor and UC Fruit Firmness testers, and the smaller and more convenient Effegi penetrometer (Figure 10.6). More sophisticated, but not necessarily more effective, tests can be carried out using machines such as the Instron Universal Testing Instrument. However, machines of this type are mostly only of value for experimental studies. While the comparatively inexpensive penetrometers do not necessarily give the same numerical value if used on the same produce, each instrument will give reproducible values for the purpose of comparative evaluation. Therefore, it is necessary to specify the instrument used when reporting pressure test values or attempting to set standards.

### Electrical characteristics

Changes in resistance or capacitance as a result of changes in the concentrations of dissolved electrolytes of the flesh during maturation have been demonstrated. However, while these measurements can be useful in the laboratory, they appear to be of little practical value.

### Chemical measurements

Measuring the chemical characteristics of produce is an obvious approach to determining maturity, particularly as they can often be directly related to taste (e.g. sweetness, sourness) of produce. The conversion of starch to sugar during maturation is a simple test for the maturity of some apple varieties. It is based on the reaction between starch and iodine to produce a blue or purple colour. The intensity of

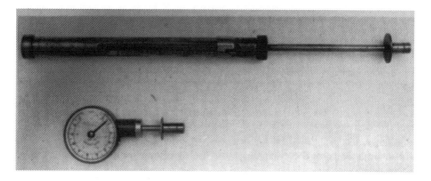

**Figure 10.6**
Effegi (bottom) and Magness–Taylor (top) fruit pressure testers.

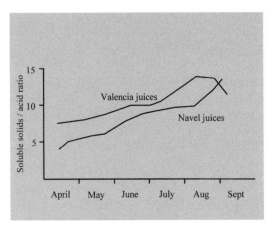

**Figure 10.7**
Variation in soluble solids/acid ratio in juice from Washington Navel and Valencia late oranges grown on trifoliata rootstocks. SOURCE Adapted from B.V. Chandler, K.J. Nicol and C. von Biedermann (1976) Factors controlling the accumulation of limonin and soluble constituents within orange fruits, *J. Sci. Food Agric.* 27: 866–76.

the colour indicates the amount of starch remaining in the fruit. This test can also demonstrate the disappearance of starch from the pulp of ripening bananas and the petals of ageing rose flowers. Sugar is usually a major component of soluble solids in cell sap and can be measured directly by chemical means. However, it is much easier and just as useful to measure total soluble solids in extracted juice with a refractometer or hydrometer. These devices operate on the basis of the refraction of light in its passage through a small sample of juice containing sugars and variation in the density of juice according to its sugar content, respectively. Maturity standards for melons, grapes and citrus are often based on the level of soluble solids.

Acidity (titratable acidity) is readily determined from a sample of extracted juice by titration with an alkaline solution (normally 0.1 N NaOH) until a pH indicator such as phenolphthalein changes colour or until a specific pH is reached (commonly 8.1). Loss of acidity during maturation and ripening is often rapid. The sugar/acid or total soluble solids/acid ratio is often better related to the fruit's palatability than either sugar or acid levels alone. Maturity standards for citrus fruits are commonly expressed as total soluble solids/acid ratios, both measurements being on a weight-for-weight basis. In Figure 10.7, the increase in the sugar/acid ratio is shown for Valencia and Washington Navel orange fruit maturing on the tree. The minimum soluble solids content is often included in specifications of fruit for processing (commonly for citrus and pineapple) and for grapes for drying or juice production. Titratable acidity and pH are not directly related as the latter depends on the concentration of free hydrogen ions and the buffering capacity of the extracted juice. Nonetheless, pH is a convenient measure that can be easily made with an inexpensive pH meter.

For example, pH is widely used in the wine industry as a quality measure for expressed grape juice.

Dry matter content, which tends to increase as certain fruit mature, can be used for products in which there is a large increase in the amount of starch or sugar as the fruit mature (e.g. kiwifruit and mangoes). The percentage of dry matter (i.e. dry weight / fresh weight x 100) can be conveniently and rapidly determined using a microwave oven to dry the material. Although dry matter content is most widely used for avocados, oil content can also be used as a maturity index for this compositionally unusual fruit.

## Climacteric behaviour

Commerical maturity can be related to the rise in the respiration rate and/or ethylene production in climacteric fruits. For longest storage and optimum eating quality, apples and pears should generally be picked just before the climacteric rise. If picked too early in the preclimacteric period, the quality when ripe will be poor. In practice, the appropriate point on the respiration curve must usually be related to some other maturity characteristic that can be readily determined in the field (see above). Nevertheless, inexpensive instruments for measuring carbon dioxide (respiration) and ethylene have been developed and used to measure internal gas concentrations, such as ethylene concentrations in the hollow core of apples. In the case of strip (once over) picking, harvesting might be commenced when a certain small proportion of the fruit has commenced ripening.

The term 'green-life' has been coined to describe the degree of physiological immaturity at the time of harvest, and refers to the time elapsing before fruit begin to ripen. Relatively less mature fruit (e.g. avocados, bananas, mangoes) have a longer green life than more mature fruit. This concept is useful for expressing potential postharvest life.

## Novel technologies for determining maturity and/or defects

Meaningful determination of maturity is inherently difficult, not the least because maturity has no discrete beginning or end, being part of the continuing development of the plant or organ. Thus, researchers are continually looking for better or improved ways to determine maturity. Similarly, the search for cost-effective methods of detecting defects, particularly internal defects, is continuing. Some novel technologies that may in time find practical application for maturity determination or defect detection include chlorophyll fluorescence, computer-aided X-ray tomography, proton magnetic resonance imaging, molecular probes and volatile detectors.

Chlorophyll fluorescence spectrophotometers can be used to detect chloroplast disruption (e.g. damage to part of the light-capturing system termed 'photosystem II' and chlorophyll loss) associated with stress (e.g. chilling, high temperature) or senescence. X-ray and magnetic resonance systems are also non-destructive technologies, and are based on detecting differences in tissue density or proton mobility in tissue, respectively. Both techniques can, for example, reveal the commencement of ripening in the inner mesocarp and/or internal cavities associated with the activity of fruit fly larvae in mango fruit. Like common measures of visible colour (e.g. colour meters, camera colour image analysers), X-rays and magnetic resonance (radio frequencies) use different parts of the electromagnetic spectrum (Figure 10.8). Near infrared spectroscopy uses yet another portion of this spectrum, and is finding increasing application as a means for non-destructively determining sugar content in relation to harvest maturity of crops such as melons. Molecular probes that bind to mRNA or proteins might be used to detect early biochemical changes during ripening (e.g. synthesis of ACC synthase or ACC oxidase). Finally, solid state broadband (i.e. chemical class, such as aldehydes) or biosensors may be applied to 'sniff' produce in order to detect the presence of volatiles associated with ripening or disorders such as decay. Although many of these and other new technologies are not in commercial use today (e.g. terahertz imaging), their investigation

**Figure 10.8** Electromagnetic spectrum illustrating the wavelengths of radiation that can be used for the non-destructive detection of changes in the structure and composition of plant tissues.

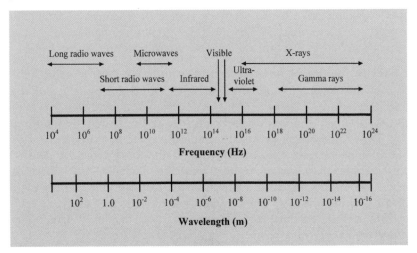

is continuing in view of the increasingly rapid advancement and change in science and technology.

The increasing contribution of computer-interfaced robotics merits singling out for mention, particularly in connection with modern sensing technologies. A current example may be seen in melon packing lines in France, where cores of mesocarp tissue are mechanically removed and tested for soluble solids to determine to which packing station individual melons should be directed.

## Postproduction quality evaluation for ornamentals

Postharvest quality testing of ornamentals also merits particular attention. Many people readily associate postharvest with harvested fruit and vegetables, but not so readily with postproduction handling of ornamentals, especially whole plants. This misconception applies equally to vegetable and ornamental seedlings, propagules such as bulbs and corms, cuttings and tissue cultures, and advanced nursery stock. Nonetheless, there is increasing awareness of the need to define and test postproduction quality of ornamentals. In Europe in particular, the ornamentals industry undertakes postproduction quality evaluation, largely because the industry there is extremely competitive and because consumers are highly discerning. The Netherlands has traditionally led the way in applied quality testing. Inspectors sponsored by the major flower auctions in Holland assess the quality of every consignment received. These same auctions also operate laboratories where specific quality-orientated tests are conducted (e.g. bacterial counts in stem ends, silver analysis for STS-treated material). Further, these laboratories conduct vase life and quality loss assessments on cut flowers and pot plants, respectively. Typical vase life assessment room conditions are 20°C and 50–70 per cent RH under an artificial fluorescent light regime of 10–20 $\mu$mol/m$^2$s at flower level for 12 hours per day. These conditions simulate the indoor home environment. Typical test conditions for pot plants are simulated transport (4 days, 16°C, 90–95 per cent RH) followed by shelf life evaluation (14 days, 20°C, 55–60 per cent RH, 8 $\mu$mol/m$^2$s for 12 hours per day). Largely subjective evalutions (i.e. based on visual appeal) are undertaken by trained professional staff employed by the auctions. Increasingly, growers and private consultants on behalf of growers or other interested parties also conduct quality evaluation. Many quality loss assessments only provide feedback after the particular consignment has been sold to consumers. With this limitation in mind, chlorophyll fluorescence parameters have been related to postproduction perfomance in an effort to develop a rapid and reliable predictive assessment of quality.

# MANAGEMENT OF QUALITY
·······

The gradual breakdown of barriers to free international trade has led to the concept of the 'global village', in which fresh produce is able to move freely from country to country based on seasonal supply and climatic conditions. Under these conditions, domestic markets in many countries are facing increased competition from imported products. Global trade is not new in that some tropical fruits such as the banana have been produced for many years in large quantities in areas like the Carribean and transported by sea to Europe and North America. Similarly large volumes of cut flowers are grown in countries such as Colombia and air freighted to Europe and North America, while apples grown in the temperate areas of the world are exported to tropical countries. The freeing of international trade has led to a much stronger emphasis on quality assurance so that buyers can rely on certain standards being maintained. There is no universal adopted set of quality standards for any given commodity, although Codex Alimentarius, Commission of WHO/FAO has grade standards used in the European Union (EU) and member countries of the Organisation for Economic and Cooperative Development (OECD). Each market has its own criteria for home consumption and for export depending on local circumstances, but generally only the higher quality lines are exported because of the longer time the produce has to survive before consumption. For many years there has been a requirement for fresh produce moving between countries to be guaranteed free of certain pests and diseases. The governments of the exporting countries are responsible for providing an inspection service and for issuing the necessary phytosanitary certificates required by importing countries, as well ensuring that imported produce does not breach domestic quarantine requirements. Recently there has been a heightened concern about food safety because of well publicised accounts of quite large numbers of people becoming ill from food-borne diseases. Consumers expect a guarantee that fresh produce does not contain harmful levels of pesticides and is free of microorganisms that cause human diseases.

Until comparatively recently, regulatory systems for quality were administered in Australia by either state or federal government agencies. The regulations were designed to maintain a base industry standard and to protect consumers from unscrupulous operators who tried to capture early season markets by harvesting produce prematurely or forwarding otherwise unsound products to market. These quality inspections were conducted at the point of despatch or at the terminal market. A factor of increasing concern to governments was the cost of providing these services. The outcome of these developments has been the withdrawal of government-funded quality inspection services, except in relation to quarantine matters and to meet the statutory requirements of other countries. The onus for the consistent delivery of quality products now rests squarely with the growers, shippers, transport operators, wholesale traders and retailers. The larger retailers in many countries are driving the development of quality assurance systems since they are usually the final point of contact with consumers.

Since the early 1990s, the horticultural industries have started to move towards a holistic system in which quality is managed along the whole distribution chain, from the farm or orchard to the final point of sale. The objective is to provide customers with a product in a condition in which he or she expects to receive it, with an assurance that the product has not been subject to a process that is likely to present any risk to health or to the satisfaction of the consumer. To do this it is necessary to monitor and prevent quality problems as early as possible in the production process rather than rely on endpoint and reactive inspection at a later stage during distribution, when the product has become more valuable. This awareness and assurance of quality is especially important when products are being shipped long distances and for long periods to overseas markets. Initially, the implementation of Quality Management Systems may entail a considerable cost, but ultimately a well managed system will reduce costs by preventing quality problems, and the supplier will gain a marketing advantage by providing consumers with the confidence that the product consistently meets their specifications.

## Quality systems

The advent of Quality Management Systems in manufacturing after World War II has led to the introduction of a plethora of terms. Total Quality Management (TQM), Quality Assurance (QA) and Quality Control (QC) refer to concepts or approaches to quality management. The application of these concepts requires the development of written descriptions of the necessary standards (the tools). The manufacturing industries have adopted a number of international quality standards (and Australian equivalents) that provide a basis for implementation and compliance with the requirements. The most common standard is AS 3902/ISO 9002, where AS refers to an Australian standard and ISO to the International Standards Organisation. The food processing industry has for many years utilised HACCP (hazard analysis and critical control points) as the basis for producing safe food.

TQM originated from the reconstruction of Japan after World War II. It is not a regulatory system but a concept in which all participants are expected to strive for continuous improvement. The EU and Australia have tended towards the adoption of a QA system based on a framework of practices and procedures, commonly ISO standards, together with a set of documented product standards or specifications. The ISO 9000 series has also been the basis of efforts to implement QA in several Australian horticultural industries, but the rate of adoption has been slow.

The Australian Horticultural Corporation (AHC), a statutory organisation established by the Australian Government, implemented the "Australian Horticulture Quality Certification Scheme" for export products. This scheme provides an additional accreditation for those companies that are already implementing ISO 9000. Export packinghouses working with the guides developed by the AHC have gained AQIS (Australian Quarantine Inspection Service) Certification Assurance, which enables them to export product without the necessity for routine inspection of each consignment. The AHC has developed a training program to help companies to begin the QA process. A comparatively recent development has been the incorporation of HACCP into documented QA systems for fresh produce. An impetus for this development has been the advent of minimally processed fruit and vegetables. The SQF 2000™ (Safe Quality Food) code developed by Agriculture Western Australia — Western Australian Government Department — is a voluntary system built around HACCP and designed to assist primary producers and small food manufacturers. In addition to SQF 2000, several other HACCP-based quality systems are being used by different sectors of the horticultural industries.

## Implementation of a Quality Management System

A Quality Management System encompasses the organisational responsibilities, procedures, activities, capabilities and resources that aim to ensure that products, processes or services will satisfy stated or implied needs. Typically a Quality Management System involves two distinct activities, both of which require detailed documentation.

◆ Quality Assurance (QA) embraces all those planned and systematic actions necessary to provide adequate confidence that the goods and services will satisfy given requirements.

◆ Quality Control (QC) determines the acceptability of a product or service by inspecting against a standard or specification, either during or after production/delivery, to assess if the required standard has been met.

The recognition of the role of HACCP in the postharvest handling of fresh produce has led to the incorporation of procedures used for many years by the food processing industry. HACCP requires detailed documented knowledge of the system under consideration. There are seven principles that characterise a HACCP-based system.

1. Identify and assess all hazards.
2. Identify the critical control points.
3. Identify the critical limits.
4. Establish the monitoring procedures.
5. Establish corrective actions.
6. Establish a record-keeping system.
7. Establish verification procedures.

A separate HACCP system is required for each product or class of products because the production systems and critical control points will differ. An example of a hazard is the misuse of a pesticide that could result in residues above the approved maximum residue limit (MRL) (Table 10.2).

The full implementation of a Quality Management System within an organisation requires total and continuing commitment from all personnel. An extensive process of training and implementation is required, and independent auditing is necessary to gain external certification of the system. A starting point for all systems is the development of detailed product descriptions and languages in conjunction with customers. These product descriptions and languages should include a set of illustrations, preferably in colour. Traditionally, the earlier government-administered quality regulations included destructive sampling of produce to ensure that they met minimum internal quality standards. Examples included firmness measurements in apples ex-storage, brix/acid ratios in early table grapes and oranges, and dry matter measurements in avocados. These measurements may be included in current product descriptions/specifications.

## Future developments

Quality Management Systems are evolving and new schemes will undoubtedly emerge. The driving forces are reliability, satisfaction and safety on the part of consumers, ensuring consistent market share and viable economic returns on the part of retailers and growers. In keeping with the globalisation of trade in horticultural products, the participants will continue to strive for Quality Management Systems that will enable them to maintain existing market share and give them access to new markets, and that are flexible in terms of catering for the range of preferences expressed by different communities. Postharvest handling now requires a full systems approach involving both the technology and recognition of human roles. For many years researchers have investigated

TABLE 10.2 IN THE PRODUCTION OF MANY HORTICULTURAL CROPS, A POTENTIAL HAZARD IS THE CONTAMINATION OF HARVEST PRODUCE WITH CHEMICAL RESIDUES IN EXCESS OF THE APPROVED MAXIMUM RESIDUE LEVEL (MRL). THIS EXAMPLE ILLUSTRATES THE APPLICATION OF THE SEVEN PRINCIPLES OF HACCP.

| HAZARD | CRITICAL CONTROL POINT | CAUSE OF HAZARD | PREVENTATIVE MEASURE | CRITICAL LIMIT | MONITORING | CORRECTIVE ACTION |
|---|---|---|---|---|---|---|
| Pesticide residue application | Yes | Insufficient harvest interval | Adhere to withholding period prior to harvest | Withholding period stated | Application and harvest dates | Spray records |
| Exceeding MRL application | Yes | Incorrect application rate of chemical | Apply chemical at recommended rate | Recommended rate | Spray records | Spray records |

mechanised techniques for the nondestructive testing of quality attributes. While advanced optical systems are now widely used for sorting produce for size, colour and surface blemishes, there is still some way to go before internal quality attributes can be measured economically using nondestructive techniques and with the required accuracy and speed.

# FURTHER READING

Abbott, J.A., A.E. Watada and D.R. Massie (1976) Effegi, Magness–Taylor and Instron fruit pressure testing devices for apples, peaches and nectarines, *J. Am. Soc. Hortic. Sci.* 101: 698–700.

Agriculture Western Australia (1995) *SQF 2000*[TM] *Quality Code: 1995*, AGWEST Quality Management, Agriculture Western Australia, South Perth.

Anon. (1985) Vegetable products, processed; solids (total) in processed tomato products: Microwave oven drying method, *J. Assoc. Official Analytical Chemists* 68: 390–91.

Askar, A. and H. Treptow (1993) *Quality Assurance in tropical fruit processing*, Springer-Verlag, London.

Australian Horticultural Corporation, (1992) *A guide to quality management and export requirements in horticulture*, Australian Horticultural Corporation, Sydney.

—— (1993) *Guide to quality management for apples*, Australian Horticultural Corporation, Sydney:

Beattie, B.B. and L. Revelant (eds) (1992) *Guide to quality management in the citrus industry*, Australian Horticultural Corporation, Sydney.

Chen, P., M.J. McCarthy and R. Kauten (1989) NMR for internal quality evaluation of fruits and vegetables, *Trans. ASAE* 32: 1747–53.

Chen, P. and Z. Sun (1991) A review of non-destructive methods for quality evaluation and sorting of agricultural products, *J. Agric. Engineer. Res.* 49: 85–98.

Gacula, M.C. and J. Singh (1984) *Statistical methods in food and consumer research*, Academic Press, Orlando, FL.

Hall, L.D. and T.A. Carpenter (1992) Magnetic resonance imaging: A new window into industrial processing, 10: 713–21.

Kapelis, J.C. (ed.) (1987) *Objective methods in food quality assessment*, CRC Press, Boca Raton, FL.

Kader, A.A. (ed.) (1992) *Postharvest technology of horticultural crops* (2nd ed.), University of California, Berkeley, CA. Publication no. 3311.

Kavanagh, E.E. and W.B. McGlasson (1983) Determination of sensory quality in fresh market tomatoes, *CSIRO Food Res. Quart.* 43: 81–89.

McGuire, R.G. (1992) Reporting of objective color measurements, *HortSci.* 27: 1254–55.

Marchant, J.A. (1990) Computer vision for produce inspection, *Postharv. News Inform.* 1: 19–22.

Meilgaard, M., G.V. Civille and B.T. Carr (1991) *Sensory evaluation techniques* (2nd ed.), CRC Press, Boca Raton, FL.

Organisation for Economic Co-operation and Development (1976) *International standardisation of fruit and vegetables: Apples and pears, tomatoes, citrus fruit, shelling peas, beans and carrots*, OECD, Paris.

Pal, D., S. Sachdeva and S. Singh (1995) Methods for determination of sensory quality of foods: A critical appraisal, *J. Food Sci. Technol.* 32: 357–67.

Priest, K.L. and E.C. Lougheed (1981) *Evaluating apple maturity — Using the starch-iodine test*, Ministry of Agriculture and Food, Ontario.

Rechnitz, G.A. and M.Y. Ho (1990) Biosensors based on cell and tissue material, *J. Biotech.* 15: 201–18.

Schmidt, S.J., X. Sun and J.B. Litchfield (1996) Applications of magnetic resonance imaging in food science, *Crit. Rev. Food Sci. Nutr.* 36: 357–85.

Scott, K.J., S.A. Spraggon and R.L. McBride (1986) Two new maturity tests for kiwifruit, *CSIRO Food Res. Quart.* 46: 25–31.

Shewfelt, R.L. and S.E. Prussia (eds.) (1993) *Postharvest handling — A systems approach*, Academic Press, San Diego, CA.

Silliker, J.H., A.C. Baird-Parker, F.L. Bryan, J.H.B. Christian, T.A. Roberts and R.B. Tompkin (eds) (1988) *HACCP in microbiological safety and quality*, Blackwell Scientific Publications, Boston, MA.

Smillie, R.M., S.E. Hetherington, R. Nott, G.R. Chaplin and N.L. Wade (1987) Applications of chlorophyll fluorescence to the postharvest physiology and storage of mango and banana fruit and the chilling tolerance of mango cultivars, *ASEAN Food J.* 3: 55–9.

Stone, H. and J.L. Sidel (1985) *Sensory evaluation practices*, Academic Press, Orlando, FL.

Sumner, J.A. (1995) *A guide to food quality assurance*, Food Technology Department, Barton College of TAFE, Moorabbin, Vic.

Vadgama, P. (1990) Biosensors: Adaptation for practical use, *Sensors and Actuators* B1: 1–7.

van Kooten, O., M. Mensink, E. Otma and W. van Doorn (1991) Determination of the physiological state of potted plants and cut flowers by modulated chlorophyll fluorescence, *Acta Hortic.* 298: 83–91.

Voss, D.H. (1992) Relating colorimeter measurement of plant color to the Royal Horticultural Society color chart, *HortScience* 27: 1256–60.

# 11
# PREPARATION FOR MARKET

This chapter on preparation for market covers postharvest handling and treatments from harvest until produce leaves the packinghouse. The main operational areas addressed are harvesting, postharvest treatments, grading and packaging. Control of diseases (Chapters 8 and 9) and evaluation and management of quality through to the retail level have been discussed earlier (Chapter 10), while storage recommendations are discussed later (Chapter 13).

## HARVESTING
......

Harvesting involves removing produce from its production environment. In the case of potted plants, harvesting is not a particularly traumatic event. On the other hand, the harvest operation may involve severing vegetables and cut flowers from their source of water, organic nutrients such as sugar, inorganic nutrients such as calcium, and growth regulators such as cytokinins and abscisic acid, in addition to eliciting wound responses such as ethylene production and increased respiration in the tissue. The response of mature fruit may be considered somewhat intermediate between pot plants and cut vegetative material, since fruit are destined for natural separation (abscission) and tend to have stored carbohydrate reserves and relatively low respiration and transpiration rates.

Harvest operations are generally carried out in the morning, when the heat load is low. Around the clock mechanical harvesting is carried out for some processing crops in order to make the best use of expensive machinery and to meet factory processing schedules. Night harvesting is used in one production area in Australia for fresh market melons. With some crops it is prudent to wait before harvesting until the early morning dew is gone and some fruit turgor is lost. This is particularly so for oranges; when fully turgid, the oil cells of the flavedo of oranges readily rupture under applied pressure (e.g. from other fruit in a bin) and spill their toxic contents. The oil can cause skin discolouration and injury, the visible symptoms of which are known as oleocellosis.

Similarly, certain mango varieties such as Kensington Pride are more prone to spurt and ooze sap when fully turgid. Mango laticifer sap stains and is caustic, both to the fruit and to operators. In some rapidly respiring crops with little or no carbohydrate reserves (e.g. cut flower crops) there may be a case for harvesting late in the day when carbohydrate levels are highest. However, in the absence of significant storage capacity, it is probable that the advantages of an early morning harvest predominate.

Harvest mechanisms include breaking off (e.g. twisting off pineapples by hand), cutting (e.g. snipping off mandarins and table grapes with secateurs) and shaking (e.g. vibrating almonds with mechanical tree trunk shakers). The mechanism chosen largely depends on how firmly the organs are attached and the degree of damage that can be tolerated, in addition to economic considerations. For fruit such as olives, weakening of the attachment of fruit following application of a chemical loosening agent (viz. ethephon, which breaks down to liberate ethylene) facilitates their harvest. For fruit on high trees, such as avocados, picking poles fitted with a cutter device that can sever and grip the fruit may be used. The motorised cherry picker, with extendable arms that can support an operator and a picking bag, is an extremely useful but expensive picking aid.

The harvest operation may be wholly by hand or by machine, or by some process that combines both approaches. Hand harvesting predominates for fresh market produce, particularly for produce susceptible to physical injury (e.g. apples, roses). On the other hand, at least in regions where labour costs are high, machine harvesting is popular for processing crops (e.g. peas and beans for freezing, peaches for canning, grapes for winemaking). Similarly, mechanical harvesting is used for robust, low unit value ground crops, such as potatoes and onions. Combination processes make use of mechanical aids, such as conveyors used to move hand-harvested pineapples and melons from the rows of plants to bins mounted on a trailer moving along parallel vehicle access aisles. Novel harvest aids include platforms on which people lie or sit to gather strawberries, and 'flying foxes' (overhead ropeways) to convey heavy banana bunches into the packing house.

In short, harvest processes and aids are as many and varied as horticultural crops themselves. In some production systems, harvesting, packing and even cooling can all be undertaken in the field. For example, lettuce grown on very large farms in California may be cut and elevated to a packing platform for cleaning, trimming and packing. The packed cartons may then be cooled in mobile vacuum coolers.

An important precaution at harvest is to avoid contamination of produce with pathogens. In this respect, practices such as standing mangoes upside down on the ground to allow the sap to drain should be discouraged. Another important precaution is to avoid physical injury. For example, deceleration shutes can be used in transferring fruit from picking bags to bins; picking bins and farm trailers can be lined with padding; and farm roads can be kept in a well maintained (i.e. smooth) condition. Harvested produce should be kept shaded, either by natural means (e.g. tree canopy) or artificial means (e.g. tarpaulins).

# POSTHARVEST TREATMENTS

A plethora of postharvest treatments is applied to horticultural crops, either to maintain quality or to improve visual appeal. Handling of harvested produce during each and every postharvest treatment process must be done with care to avoid mechanical damage. For instance, large drop heights, friction areas and sharp objects should not be encountered on the treatment and packing line.

## *Washing*

Washing may be important to remove sap (e.g. mangoes), dirt (e.g. carrots) and debris (e.g. bananas). Clean water is essential as fungal and bacterial levels may otherwise build up. A series of two or three washes may be beneficial, possibly with an approved disinfectant treatment (e.g. ultraviolet light, ozone, chlorination) applied to the last wash. Soap and other chemicals such as calcium hydroxide can also be added to washing water to facilitate the process.

## *Waxing*

Commodities that have a waxy skin tend to lose water slowly. This observation has led to the application of wax to certain commodities that shrivel rapidly and lose consumer appeal during storage and marketing (e.g. defuzzed peaches). In addition to reducing water loss, waxes are also applied to improve the appearance of produce to the consumer (e.g. shiny apples). The rate of water loss can be reduced by 30–50 per cent under commercial conditions, particularly if the stem scar and other injuries are coated with wax. Citrus fruit are commonly waxed because washing can remove much of the natural wax from the peel, thereby exacerbating shrivelling and loss of appearance. Many other commodities — such as cucumbers, tomatoes, passionfruit, peppers, bananas, apples and some root crops — are also waxed to reduce weight loss and to increase their consumer appeal.

Most waxes in commercial use are proprietary formulations. These formulations may contain a mixture of waxes derived from plant and/or petroleum sources. Many have been based on a combination of paraffin wax, which gives good control of water loss but a poor lustre to the produce, and carnauba wax, which imparts an attractive lustre to the produce but provides poorer control of water loss. In addition, formulations containing polyethylene, synthetic resin materials, sugars and sugar derivatives, chitosans and emulsifying and wetting agents have been used. The wax formulation may be used as a carrier of fungicides or of inhibitors of senescence, superficial scald or sprouting. Waxes are brushed, sprayed, fogged or foamed (Figure 9.3 in the colour plate section) onto produce, or produce is conveyed through a tank of wax emulsion. The wax film must be thin or else gas exchange may be overly hindered, causing anaerobiosis and associated quality loss such as the production of off flavours. Following the application of wax, produce is generally dried and polished.

## Curing

Underground storage organs such as potato and sweet potato tubers tend to have poorly developed cuticles. Thus, they are relatively susceptible to mechanical wounding during harvesting and handling, and to postharvest water loss and decay. These problems can be minimised by the process of 'curing' at intermediate to high temperatures and at high humidity. For example, sweet potatoes can be cured at around 30°C and 90–95 per cent RH for 4–7 days prior to storage. Similarly, potatoes may be cured at 10–15°C and about 95 per cent RH for a period of 10–14 days. During the curing period, a surface layer of protective suberised wound periderm tissue (Chapter 2) is formed over the product, especially at wound sites. Although most rapid periderm formation on potatoes occurs at about 21°C, the risk of decay at this comparatively high temperature is unacceptable. Curing treatments may also facilitate wound healing for certain fruit, such as citrus.

## Plant growth regulators

The five major types of plant growth regulators are auxins, gibberellins (GA), cytokinins, abscisic acid (ABA) and ethylene. These phytohormones control many plant functions and biosynthetic pathways, and have been ascribed a regulatory role in all aspects of plant development from germination and primordia formation to senescence. Postharvest information on the effect of the first four types of compounds is not as extensive as for ethylene. However, they are all worthy of ongoing study because they regulate physiological processes at extremely low concentrations. Since phytohormones occur naturally, they might be expected to receive consumer acceptance as potential food additives. Nonetheless, testing and approval by health authorities is necessary.

Despite their potency and natural occurrence, few postharvest applications of plant growth regulators, except ethylene (viz. controlled abscission, degreening and ripening; see below), have been implemented by industry. Two commercial examples are:

◆ retardation of button senescence on citrus fruit by 2,4-dichloro-phenoxyacetic acid (2,4-D) in order to prevent development of stem-end rots; and

◆ retardation of the senescence of the peel of citrus fruit by gibberellic acid ($GA_3$).

It has been shown in laboratory studies that cytokinins, notably the synthetic cytokinin benzyladenine (BA), significantly delays the loss of chlorophyll from, and the general senescence of, many green leafy materials (e.g. broccoli florets, alstroemeria leaves). Exogenously applied $GA_3$ can reduce ethylene production by some cut flowers, and ABA can be applied to help maintain the water balance of cut flowers and pot plants through induced stomatal closure.

### Sprout inhibitors

The buds of potatoes and onions enter a dormant state at maturity by which process the plant can survive periods unfavourable to growth. These vegetables are normally harvested at this time, and can be stored for many months under correct conditions for either retail marketing or processing. In potatoes, the duration of postharvest dormancy (rest period) is influenced by preharvest factors, maturity and variety, but generally not by the temperature of storage. Once the rest period ends, the rate of sprouting depends on temperature. Sprouting of potatoes rarely occurs below 4°C, but storage at these temperatures is impractical due to the conversion of starch to sugars (Chapter 4). At temperatures greater than 4°C, sprouting is a problem during storage periods greater than 2–3 months.

The application of several chemicals and ionising radiation (see later this chapter) effectively suppresses sprouting in potatoes and onions during storage at higher temperatures. Maleic hydrazide (MH), nonyl alcohol, 3-chloroisopropyl-N-phenylcarbamate (CIPC), isopropyl-phenylcarbamate (IPPC), methyl naphthaleneacetic acid (MENA) and 2,3,4,6-tetrachloronitrobenzene (TCNB) are commercial sprout inhibitors. However, legal restrictions on the usage and permitted residues of these compounds vary among countries. CIPC is a strong sprout inhibitor and is probably the most widely used of the chemical sprout inhibitors for the storage of potatoes. CIPC may be applied as a dust, water dip, vapour or aerosol. Like many of the other chemical sprout inhibitors, CIPC interferes with periderm formation, and thus should be applied after curing (see above). Sprouting of onions during long-term storage is effectively prevented by MH applied several weeks before harvest.

## Disinfestation

Some insect species, such as the tephritid fruit flies that infest a broad range of horticultural fruit crops, can seriously disrupt produce trade between countries. These insects may also restrict the movement of produce within a country. Table 11.1 lists the more important species, their distribution and hosts. Importing countries often place a quarantine barrier against produce originating from an area in which the insect species of concern is known to occur. In order to market produce, exporting nations with quarantine insect pests must develop an effective disinfestation treatment that satisfies the importing country, virtually by killing each of the egg, larva, pupa and adult development stages. However, the disinfestation treatment must not harm either the produce or the consumer, and should be economical to apply.

Disinfestation protocols employing chemical (e.g. ethylene dibromide [EDB]) or physical (e.g. vapour heat) treatments have been developed to kill insects infesting horticultural produce. Fumigation with gaseous sterilants is the most important technique for disinfesting produce. However, these are falling from favour because of high mammalian toxicity (e.g. hydrogen cyanide), flammability (e.g. carbon disulphide) and damage to the atmospheric ozone layer (e.g. methyl bromide [MB]). For example, the use of EDB as a fumigant on produce for consumption in the USA has been prohibited since 1984 because of suspected carcinogenicity. Other countries have also banned its use. Similarly, MB will cease to be manufactured in the USA from the year 2000. Gaseous sterilants that have been used for general quarantine fumigation include acrylonitrile, carbon disulphide, carbon tetra-chloride, EDB, ethylene oxide, hydrogen cyanide, MB, phosphine and sulphuryl fluoride. Currently, MB is most widely used for fresh produce and phosphine for dry or stored produce.

Maximum permissible residues of chemical disinfestants are specified by law in most countries. Subject to regulations, fumigants can be applied to produce in a permanent fumigation chamber, in a temporary enclosure (e.g. under a tarpaulin) or in gas-tight rail cars and road trucks.

## TABLE 11.1 SOME INSECTS AND MITES THAT CAN BE CARRIED BY FRUIT AND VEGETABLES

| INSECT | COMMON NAME |
| --- | --- |
| *Cryptophlebia leucotreta* Meyr. | False codling moth |
| *Cydia pomonella* (L.) | Codling moth |
| | |
| *Cylas formicarius* (Fab.) | Sweet potato weevil |
| | |
| Fruit flies, such as | |
| *Anastrepha fraterculus* (Wied.) | South American fruit fly |
| | |
| *A. ludens* (Lw.) | Mexican fruit fly |
| | |
| *Ceratitis capitata* (Wied.) | Mediterranean fruit fly |
| | |
| | |
| *C. rosa* Karsch | Natal fruit fly |
| *Bactrocera ciliatus* (Lw.) | Lesser pumpkin fly |
| *Bactrocera cucurbitae* Coq. | Melon fly |
| | |
| *Bactrocera dorsalis* (Hend.) | Oriental fruit fly |
| *Bactrocera tryoni* (Frogg.) | Queensland fruit fly |
| *Rhagoletis cerasi* (L.) | Cherry fruit fly |
| *R. cingulata* (Lw.) | Cherry fruit fly |
| *Graphognathus leucoloma* (Boh.) | White fringed weevil |
| | |
| *Halotydeus destructor* (Tucker) | Red-legged earth mite |
| *Lobesia botrana* (Schiff.) | Vine moth |
| *Maruca testulalis* (Geyer) | Bean pod borer, mung moth |
| | |
| Mealybugs, such as | |
| *Planococcus citri* (Risso) | Citrus mealy bug |
| *Dysmicoccus bevipes* (Ckll.) | Pineapple mealy bug |
| | |
| *Panonychus ulmi* (Koch) | European red mite |
| | |
| *Phthorimaea operculella* (Zell.) | Potato tuber moth |
| Scale insects, such as | |
| *Aonidiella aurantii* (Maskell) | Red scale |
| *Lepidosaphes beckii* (Newm.) | Purple scale |
| *Quadraspidiotus perniciosus* (Comst.) | San José scale |
| *Sternochaetus mangiferae* (Fab.) | Mango seed weevil |

SOURCE Prepared from *distribution maps of insect pests* issued by the Commonwealth Institute

| COMMON HOSTS | APPROXIMATE DISTRIBUTION |
| --- | --- |
| Citrus, avocado, stone fruit, guava | Africa |
| Apple, pear, peach, quince, *Prunus* species, walnut | Worldwide |
| Sweet potato | Africa, Asia, Pacific Islands, North America, South America |
| Peach, guava, citrus, *Spondias* species, *Eugenia* species | South America, West Indies, Central America |
| Citrus, other tropical and subtropical fruits | Central America, Mexico |
| Deciduous and subtropical fruits, especially peach and citrus | Southern Europe, Africa, Central America, South America, Western Australia, Hawaii |
| Many deciduous and subtropical fruits | Africa |
| Cucurbits | Africa, India, Pakistan, Bangladesh |
| Cucurbits, tomato | Asia, Hawaii, Papua New Guinea, Africa |
| Most fleshy fruits or vegetables | Asia, Hawaii |
| Many deciduous and subtropical fruits | Australia, Pacific Islands |
| Cherry, *Lonicera* species | Europe |
| Wild and cultivated cherry, *Prunus* species | North America |
| Root vegetables | South Africa, Australia, New Zealand, USA, South America |
| Leafy vegetables | Australia, New Zealand, Africa |
| Grape | Europe, Japan, Africa |
| Legumes | Africa, Asia, Australia, Central America, South America, Pacific Islands |
| Citrus, grape | Worldwide |
| Pineapple | Africa, Asia, Australia, Pacific Islands, South America |
| Apple and other deciduous fruits | Europe, Africa, Asia, Australia, New Zealand, North America, South America |
| Potato, tomato, egg-plant | Worldwide |
| Citrus | Worldwide |
| Citrus | Worldwide |
| Deciduous fruits | Worldwide |
| Mango | Africa, Asia, Australia |

:omology, London.

**195**

**Figure 11.2**
Experimental vapour heat treatment apparatus. Heated saturated air is forced through boxes of produce contained in the chamber in the foreground. (Courtesy of Dr M. Brown, CSIRO Division of Horticulture.)

The permanent chamber is the safest and most reliable fumigation option. Fumigation under tarpaulins offers flexibility at the dockside or railway yard, and if correctly carried out is as effective as chamber fumigation. Complete operating details applicable in the USA for disinfesting produce by both chemical and physical procedures are found in the *Plant Protection and Quarantine Treatment Manual* prepared by the Animal and Plant Health Inspection Service of the US Department of Agriculture. Other countries have similar procedures that should be carefully followed in order to obtain safe and effective treatment.

High carbon dioxide and/or low oxygen atmospheres have been investigated as alternatives to the chemical treatments mentioned above.

These have had limited success. For example, high carbon dioxide is used commercially to control insects infesting dried grapes. Aerosol formulations of insecticides such as synthetic pyrethrins or dichlorvos are used to disinfest ornamentals such as cut flowers. However, unlike gases, aerosols comprise small droplets that settle out of the air onto surfaces. Consequently, insects that are shielded from the aerosol droplets are often not harmed. These insects may be protected by an overhead canopy of stems, foliage or flowers, and by the impenetrable walls of plastic buckets. Postharvest insecticide sprays and dips are, however, often used for cut flowers and foliage and for pot plants, as well as for some fruit and vegetables. The insecticide dichlorvos has some vapour phase activity. Slow release dichlorvos-based pest strips have been included in cartons packed with flowers to effect ongoing disinfestation during export.

Minimum effective doses of chemical or physical disinfestation treatments that leave no survivors have been established through research for a range of insects and horticultural commodities (Figure 11.1). For produce that is sensitive to chemical disinfestants, alternative disinfestation procedures have been developed that mainly involve high or low temperatures. The commercial use of these treatments has also increased because of concern by authorities and consumers about chemical residues in produce.

Many insects that infest produce do not tolerate exposure to low temperatures, an observation that has led to an effective disinfestation procedure for deciduous fruits but not for tropical and subtropical fruits, which are liable to chilling injury. For instance, US quarantine authorities stipulate the following cold treatments for produce from areas infested with the Mediterranean fruit fly: 10 days at 0°C or below; 11 days at 0.6°C or below; 12 days at 1.1°C or below; 13 days at 1.7°C or below; or 16 days at 2.2°C or below. To guarantee disinfestation against both Queensland and Mediterranean fruit flies, Japanese quarantine authorities have accepted cold treatment of Australian oranges against Queensland and Mediterranean fruit flies for 16 days below 1.0°C, measured with a tolerance of 0.6°C. Similar treatments are being developed for other cold-tolerant fruits with potential for export to Japan and other countries. Some importing countries permit cold treatments to be applied during transit. Cold treatments may also be applied in combination with chemical fumigation, thus reducing the amount of fumigant required.

Produce can also be successfully disinfested by exposure to high temperature. High temperature treatments can be achieved with hot air, including vapour heat (Figure 11.2), or hot water. For instance, with

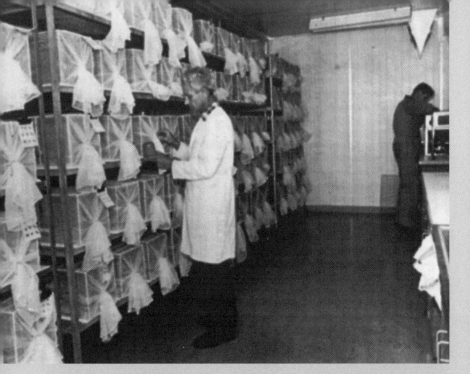

**Figure 11.1** Establishment of effective treatments for the disinfestation of produce. *Above right:* Breeding fruit flies. *Below right:* Apple being experimentally infested with light brown apple moth larvae. *Above:* Fruit flies laying their eggs in oranges. (Courtesy of W.E. Rushton, formerly CSIRO Division of Food Research.)

vapour heat treatment, the temperature of fruit such as citrus, mangoes, papaya and pineapples and some vegetables is raised to about 47°C with air saturated with water vapour, and the core temperature of the fruit is maintained at that elevated temperature for a prescribed period of time (e.g. at least 15 minutes). As with low temperature treatments, the precise high temperature treatment regime, which must be strictly complied with, are commodity, pest and method specific, and are the subject of negotiated agreement between the exporting and importing country.

## *Irradiation*

National governments have been somewhat unwilling to approve ionising radiation as a disinfestation treatment for fresh fruit and vegetables. However, the USDA has recently proclaimed schedules for irradiation at 150–250 Gy (gray) against fruit fly infestations. FAO/IAEA guidelines for quarantine security prescribe doses of 75, 150 and 300 Gy for the Queensland fruit fly, all fruit flies and any stage of any insect, respectively. In contrast to fresh produce, the preservation of processed food (e.g. dried herbs) by ionising radiation is practised in many countries.

The potential benefits of gamma or electron beam radiation in the postharvest handling of fruit and vegetables lie in both insect and disease disinfestation, and in retarding aspects of produce development, such as ripening and sprouting. Before this potential can be realised, several criteria must be met:

◆ the host must have a considerably higher tolerance to radiation than the organisms or metabolic systems causing deterioration;

◆ the required radiation treatment must be as, or more, economical than competing effective chemical and physical treatments; and

◆ the radiation treatment must be acceptable to both health authorities and consumers (i.e. the safety and wholesomeness of the irradiated produce must be demonstrated).

These three criteria may also be generalised to the evaluation of heat and cold disinfestation treatments. Disinfestation of fruit, particularly tropical fruits such as papaya and mangoes, by gamma radiation has been shown to be technically feasible. The egg phase of the life cycle of insects is the most sensitive to irradiation, followed in order by larval, pupal and adult stages. Most insects are sterilised at doses of 50–200 Gy. However, some adult moths will survive 1 kGy, although their progeny are sterile.

While no fruit is yet being commercially disinfested by irradiation, public perceptions and acceptance may change with the introduction of more and tighter restrictions on disinfestation by chemicals, including clear labelling of the products. By 1985, 30 countries had issued provisional or conditional clearances covering the irradiation of more than 50 different food items. Twenty of these clearances were for sprout inhibition of potatoes and onions. Japan makes considerable use of irradiation of potatoes to suppress sprouting. Sprouting in potatoes and onions during long-term storage is inhibited by 20–150 Gy. This dose range has little effect on other aspects of potato and onion quality, such as sugar level, rates of decay and water loss, texture and flavour. Irradiation of potatoes and onions is, however, more expensive than treatment with the chemical sprout inhibitors CIPC and maleic hydrazide, but is residue free.

Disinfestation and sprout inhibition are good examples of instances where the difference in sensitivity to the low doses of ionising radiation between the produce item and the organism or metabolic system to be restricted is sufficiently great for the irradiation process to be technically feasible. When produce, particularly fruit, is irradiated with intermediate levels of ionising radiation of 0.5–2 kGy to control the growth of common decay organisms, the tissues of the fruit are often severely damaged. The most common effect of ionising radiation is to soften produce, making it more susceptible to damage during subsequent handling and transport.

At present, even for those commodities that respond favourably to irradiation, cheaper and often more effective alternative disinfestation procedures are available (Table 11.2). Nonetheless, irradiation may become more attractive for certain items, such as potatoes and onions and high-cost tropical fruits for export, if the general public raises strong objections to the continued use of chemicals for disinfestation, wastage control and sprout inhibition and if the cost of irradiation can be lowered. Three joint FAO/WHO reports on the wholesomeness of irradiated foods have dispelled many early fears about induced radioactivity, loss of nutritional value or the production of harmful radiolysis products in irradiated produce. The first of these reports (WHO, 1977) recommended the acceptance of nine irradiated foods, including potatoes, onions, mushrooms, papaya and strawberries. The second report (WHO, 1981) extended the list of approvals to include, among others, mangoes, and concluded that the irradiation of any food commodity up to an overall average dose of 10 kGy presents no toxicological hazard, and hence toxicological testing of foods so treated was no longer required. These conditions were accepted by the Codex Alimentarius Commission in 1983 when drafting a general standard for irradiated foods. However, there has not yet been any marked expansion in commercial food irradiation because of the perceived reluctance of consumers to accept irradiated foods.

**TABLE 11.2  COMPARISON OF MAXIMUM TOLERABLE AND M
TECHNICAL EFFECTS ON SELECTED FRESH PRODUCE**

| PRODUCE | DESIRED TECHNICAL EFFECT | MAXIMUM TOLERABLE DOSE (KGY) ★ |
|---|---|---|
| Apple | Control of scald and brown core | 1–1.5 |
| Apricot, peach, nectarine | Inhibition of brown rot | 0.5–1 |
| Asparagus | Inhibition of growth | 0.15 |
| Avocado | Inhibition of ripening and rot | 0.25 |
| Banana | Inhibition of ripening | 0.5 |
| Lemon | Inhibition of *Penicillium* rots | 0.25 |
| Mushroom | Inhibition of stem growth and cap opening | 1 |
| Orange | Inhibition of *Penicillium* rots | 2 |
| Papaya | Disinfestation of fruit fly | 0.75–1 |
| Pear | Inhibition of ripening | 1 |
| Potato | Inhibition of sprouting | 0.2 |
| Strawberry | Inhibition of grey mould | 2 |
| Table grape | Inhibition of grey mould | 0.25–0.50 |
| Tomato | Inhibition of *Alternaria* rot | 1–1.5 |

★1 Gray = 100 rads.

SOURCE  Adapted from E.C. Maxie, N.F. Sommer and F.G. Mitchell (1971) Infeasibility of

## Superficial scald control (apples)

The superficial scald disorder of apples in cool storage is associated with the oxidation products of $\alpha$-farnesene. These products arise when the natural antioxidants in the fruit are degraded or inactivated during cool storage (see Chapter 8). The addition of various synthetic antioxidants effectively prevents the oxidation of $\alpha$-farnesene, and hence the development of scald. Commercially, scald has been reduced by a postharvest dip in either diphenylamine (0.1–0.25 per cent) or 1.2-dihydro-6-ethoxy-2.24-trimethylquinoline (ethoxyquin) (0.2–0.5 per cent). Diphenylamine may also be applied by means of impregnated wraps or in wax. These

UM RADIATION DOSE REQUIRED FOR DESIRED

| MINIMUM DOSE REQUIRED (kGY) | PHENOMENA LIMITING COMMERCIAL APPLICATION |
| --- | --- |
| 1.5 | Cheaper, more effective alternatives; tissue softening |
| 2 | Tissue softening |
| 0.05–0.1 | Economics, short season, small acreage |
| none | Cheaper, more effective alternatives; browning and softening of tissues |
| 0.30–0.35 | Cheaper, more effective alternatives |
| 1.5–2 | Severe injury to fruit at doses of 0.5 kGY or more; cheaper, more effective alternatives |
| 2 | Cheaper, more effective alternatives; no technical effect under commercial conditions |
| 2 | Cheaper, more effective alternatives; no technical effect under commercial conditions |
| 0.25 | Not economical, inadequate acreage |
| .25 | Abnormal ripening; cheaper, more effective alternatives |
| 0.08–0.15 | Cheaper, more effective alternatives |
| 2 | Cheaper, equally effective alternatives |
| | Tissue softening, severe off-flavours; cheaper, more effective alternatives |
| 3 | Abnormal ripening, tissue softening |

ing fresh fruits and vegetables, *HortScience* 6: 292–94.

antioxidants may also be brushed onto fruit. Both compounds were discovered within a few years of each other, and most countries have approved one or the other, but rarely both. This discord has caused problems in international trade in apples, where exporting countries have to use the scald treatment that is legally permitted by the importing nation. As the treatment is applied immediately after harvest, the final destination of the fruit must be known with some certainty at harvest, and some segregation of treated fruit must be maintained during storage.

Although diphenylamine and ethoxyquin adequately control scald, a single compound, approved by all countries, is desirable. For political reasons, it is unlikely that either compound will gain universal approval. Chemical companies are reluctant to pursue approval for various other compounds that have been shown to be effective scald inhibitors due to the increasingly stringent demands of health authorities.

The dilemma over scald treatments highlights the problems that can be expected more often in the future. There is a reluctance internationally to allow new chemicals as food additives. This concern particularly affects the postharvest treatment of fresh produce as food laws are mainly concerned with foods after harvest. In addition, strict maximum residue limits (MRLs) are being established for many of the agricultural pesticide chemicals. Recent research with superficial scald has focused on finding volatile compounds, such as ethanol, that are naturally produced by fruit and vegetables as possible commercial treatments that may have universal acceptance, and some success in this approach has been achieved.

## Calcium application (apples)

Since about 1960, preharvest calcium sprays have been applied commercially to apples in several countries to reduce the incidence of the physiological disorder bitter pit (dark necrotic spots on the skin) after harvest (see Chapter 8). Added calcium also reduces internal breakdown of apples in cool storage. To be effective, calcium must actually contact the fruit and be absorbed directly by it. Calcium that falls on the leaves or branches is not effectively transferred to fruit. The sprays may be applied up to six times during the growing season in order to slowly build up calcium levels in fruit. It is difficult to ensure that all fruit are wetted sufficiently by one application of calcium solution. This preharvest spray uptake problem may be overcome by dipping apples after harvest in solutions of calcium salts. Uptake can be further improved by partial pressure (reduced pressure, vacuum) or positive pressure infiltration to force calcium solution into the apple flesh.

The partial pressure infiltration technique, developed first in Australia, has been used in New Zealand on a commercial scale with considerable success. Its use in other countries has shown less promise. It seems that best results are obtained with apples that have a closed calyx so that the calcium solution is forced into the fruit through the lenticels, and is thus spread around the perimeter tissue where the disorders occur. With open calyx fruit, the uptake of solution is difficult to control as it readily enters the fruit via the calyx and excess solution accumulates in the core area, often leading to injury or rotting.

Laboratory studies with partial pressure infiltration of calcium solutions have shown that the technique can markedly retard the initiation of ripening in a number of climacteric fruit such as tomatoes, avocadoes, mangoes and pears. However, large-scale application of calcium infiltration into fruit other than apples has not been undertaken due to attendant risks of skin injury due to excess calcium uptake and rot development.

## Controlled ripening

Climacteric fruits, particularly tropical and subtropical species, are frequently harvested when less than fully ripe and then transported, often over considerable distances, to areas of consumption. These fruits are then ripened to optimum quality under controlled conditions of temperature and RH and, with some fruits, through the addition of ethylene and other gases that initiate ripening. A further advantage of controlled ripening is improved uniformity of ripening among fruit. The use of relatively high ripening temperatures may also minimise the development of rots in ripe tropical fruits. In contrast, non–climacteric fruits undergo little or no desirable change in composition after harvest, and are not harvested until fit for consumption.

A significant proportion of the world production of bananas is ripened under controlled conditions. The banana is unusual in that it can be picked over a wide range of physiological maturity ages from half-grown (thin and angular) to fully grown (full and rounded) and ripened to high quality with the aid of ethylene. It was long known that bananas could be induced to ripen by enclosing the fruit in a chamber of some kind with restricted ventilation, and in more recent times it was found that enclosing smouldering wood or charcoal with bananas hastened their ripening. It is now known that ethylene is a product of incomplete combustion of such fuels, and that it is by far the most active of the known ripening gases. Acetylene, generated by adding water to calcium carbide, also induces the ripening response, but in practice a concentration at least 100-fold higher is required.

The commercial ripening of bananas has been reduced to a routine operation, and fruit at a specified colour stage (Table 11.3) can be produced on a predetermined schedule. The effective concentration of ethylene for banana ripening is quite low (Table 11.4), and further increases in concentration give no added advantage when this concentration is maintained for the stated period. In practice, however, high concentrations are used initially because the ripening rooms are not sufficiently airtight. Bananas may be ripened by a batch process in which the chamber (Figure 11.3) is charged (the 'shot') with ethylene gas to a concentration of between 20 and 200 $\mu$L/L. The chamber has to be ventilated after the first 24 hours to prevent the carbon dioxide

## TABLE 11.3  COLOUR STAGES OF THE RIPENING CAVENDISH BANANA

| Stage | Peel colour | Approx. Starch (%) | Approx. Sugar (%) | Comment |
|---|---|---|---|---|
| 1 | Green | 20 | 0.5 | Hard, rigid; no ripening |
|  | Sprung Green | 19.5 | 1.0 | Bends slightly, ripening started |
| 2 | Green, trace of yellow | 18 | 2.5 |  |
| 3 | More green than yellow | 16 | 4.5 |  |
| 4 | More yellow than green | 13 | 7.5 |  |
| 5 | Yellow, green tip | 7 | 13.5 |  |
| 6 | Full yellow | 2.5 | 18 | Peels readily; firm ripe |
| 7 | Yellow, lightly flecked with brown | 1.5 | 19 | Fully ripe; aromatic |
| 8 | Yellow with increasing brown areas | 1.0 | 19 | Over-ripe; pulp soft and darkening, highly aromatic |

SOURCE Commonwealth Scientific and Industrial Research Organisation (1972) *Banana ripening guide*, CSIRO, Melbourne. Division of Food Research circular 8.

concentration from exceeding 1 per cent, and thus retarding ripening. If the chamber is poorly sealed it may be necessary to recharge it with ethylene after 12 hours. Fruit are removed at the desired colour stage. A more satisfactory alternative to charging the ripening chamber with an initial high concentration of ethylene is to 'trickle' ethylene into the chamber (Figure 11.4) at a rate just sufficient to maintain the concentration shown in Table 11.4. The ripening chambers should be ventilated at the rate of about one room volume each 6 hours to prevent the accumulation of carbon dioxide. In practice it is usually not necessary to install a ventilation system in rooms less than 60 m³ because they have natural air exchange (leakage) rates higher than the required minimum rate. For larger rooms, the natural air leakage rates should be

**Figure 11.3**
Controlled ripening room for bananas. This room is designed for forced air circulation.
Note the vented cartons, exhaust fan at rear and the blind that can be drawn over the
cartons to enclose the gap between the rows of pallets.

taken into account when calculating the ventilation requirement. Excess
ventilation will increase energy consumption for cooling and heating.

Traditionally, containers of bananas were stacked in a ripening room
so as to expose at least two faces to the circulating air to ensure that fruit
temperatures were even. In modern practice, the fruit are packed in
vented cartons, unitised on pallets, and fruit temperatures are controlled
by forced air circulation. Adoption of the forced air system has increased
the capacity of the rooms by about 40 per cent and enabled fruit
temperatures to be controlled more accurately. A minimum air-flow of
about 0.34 L/sec. kg of bananas is required. In Australia, the forced air
system is also commonly used in tomato ripening rooms.

Water loss can be high at ripening temperatures unless RH is
maintained at a high level. Humidity can be raised by atomising water
into the ripening chamber or simply by spraying the floor with water.
Regulation of RH during the course of ripening can be particularly

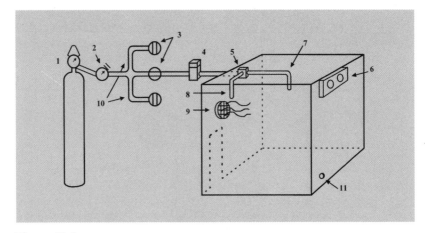

**Figure 11.4**
Trickle system for adding ethylene to a ripening room. (1) ethylene cylinder; (2) pressure regulator, with outlet pressure set on 300 kPa; (3) on–off toggle valve(s); (4) solenoid valve wired to open only when the air circulation fan is operating; (5) variable flow controller or rotameter located on the front wall of the room; (6) air circulation fan or refrigeration unit; (7) gas outlet to be located near the centre of the room in small rooms (less than 60 m³); (8) in larger rooms, the gas outlet should be located in front of the outlet from the blower fan; (9) blower fan mounted on the front wall of the room (the amount of air discharged into the room can be regulated to provide a predetermined rate of ventilation); (10) branches to other rooms if required; (11) a small exhaust porthole is provided in the rear wall of a room (usually large rooms over 60 m³) that requires mechanical ventilation.
SOURCE W.B. McGlasson, E. Kavanagh and B.B. Beattie (1986) Ripening tomatoes with ethylene, *Agfact*, NSW Department of Agriculture, H8.4.6.

important for bananas. An RH range of 85–90 per cent has been recommended for ripening to stage 2 (Table 11.3), but this should be reduced to 70–75 per cent during the later colouring stages to avoid skin splitting. Although the best skin colour may be achieved at the highest RH, commercial experience with conventional chambers not designed for forced air circulation has shown that the skin tends to be too soft and may split. If RH is too low, weight loss may be excessive, colour poorer and blemishes more pronounced. Maintenance of RH at 85–90 per cent continuously in forced air ripening rooms has proved satisfactory. Because of the high RH and temperature maintained in ripening rooms, moulds grow readily on any organic matter, including the walls of the room if they are not suitably protected. Regular cleaning with a solution of sodium hypochlorite (chlorine) followed by fumigation with

## TABLE 11.4 RIPENING CONDITIONS FOR SOME FRUITS USING ETHYLENE*

| FRUIT | TEMPERATURE (°C) | ETHYLENE (μL/L) | TREATMENT TIME (HOURS) |
|---|---|---|---|
| Avocado | 18–21 | 10 | 24–72 |
| Banana | 15–21 | 10 | 24 |
| Cantaloupe | 18–21 | Nil | |
| Honeydew melon | 18–21 | 10 | 24 |
| Kiwifruit | 18–21 | 10 | 24 |
| Mango | 29–31 | 10 | 24 |
| Papaya | 21–27 | Nil | |
| Pear | 15–18 | 10 | 24 |
| Persimmon | 18–21 | 10 | 24 |
| Tomato | 13–22 | 10 | continuously |

*Relative humidity is normally maintained at 85–90 per cent.

formaldehyde gas increases the life of ripening rooms, reduces maintenance costs and minimises fruit wastage.

The ripening of other climacteric fruit that have been harvested immature can be hastened by treatment with ethylene under controlled conditions. However, in contrast to bananas and avocados, quality will be inferior to that of fruits harvested at the mature green stage. With many of these fruit it is more important to harvest at the correct stage of maturity (e.g. at the full slip stage for cantaloupes, the first appearance of yellow colour in the blossom end of papaya and the first colour [breaker] stage of tomatoes). However, at full maturity it is only necessary to hold the temperature and RH to the specifications in Table 11.4 to achieve high quality ripened fruit. In other words, treatment with ethylene is not necessary for these fruit to ripen fully, provided they are picked when sufficiently developed. When fully developed, at least some, if not all, of the fruit will produce sufficient ethylene to effectively ripen themselves and adjacent fruit. Nonetheless, ethylene treatment will promote more uniform and rapid ripening in consignments of mixed maturities, such as strip-picked mangoes and once-over harvested tomatoes.

## Controlled degreening

The pulp of many early-season citrus cultivars becomes edible before the green colour has completely disappeared from the peel. Exposure to low temperatures during maturation is necessary for the development of an orange coloured peel. This requirement explains why the peel of citrus grown in the low altitude tropics fails to degreen completely. Furthermore, the Valencia orange cultivar is often stored on the tree for several months after ripening has been completed, and during this storage period the peel tends to regreen. Postharvest treatment with ethylene under controlled conditions hastens the loss of chlorophyll. This process is known as degreening. Batch or trickle degreening is a cosmetic treatment designed to give the fruit a ripe appearance, but does not result in significant changes in pulp composition if correctly administered. The conditions of batch degreening — 20–200 µL ethylene/L, 25–30°C and 90–95 per cent RH — are maintained for 2–3 days with regular ventilation of the chamber to prevent the build up of carbon dioxide (citrus is injured by carbon dioxide concentrations above 1 per cent). Trickle degreening, with 10 µL ethylene/L continuously metered into the room, is more rapid than batch degreening and is therefore preferable since degreening conditions accelerate deterioration and decay of citrus. Although the most rapid degreening occurs at 25–30°C, the production of peel carotenoids is greatest at 15–25°C.

The ethylene-releasing compound ethephon ripens and degreens fruit equally as well as ethylene gas. Ethephon is absorbed by fruit tissues and, when the pH exceeds 4.6, breaks down to release ethylene.

In some citrus growing areas, notably Florida in the USA, tangelos, temples and some early season oranges have not experienced enough cold weather to promote the development of a highly coloured peel. Packing sheds in these areas are legally permitted to dye the peel of these fruits under strictly controlled conditions with Citrus Red No. 2. This process is known as 'Colour Add', and can only be used on mature fruits that are not intended for processing. The dye is applied to the citrus fruit by dip or drench, and typical treatment times and temperatures are 4 minutes (maximum) at 49°C for oranges and 4 minutes (maximum) at 46°C for temples and tangelos. After treatment, the fruit are rinsed thoroughly to prevent 'bleeding' of the dye through the wax and to ensure that the Food and Drug Administration (FDA) residue tolerance of 2 ppm is not exceeded. The wax is applied after all the water is removed.

## Light (minimal) processing

There is increasing consumer demand for ready-to-use foods, including fruit and vegetables. This demand has fostered the development of the 'lightly processed' fruit and vegetable industry, which is also known as the 'minimally processed' or 'fresh cut' industry. For vegetables, the process, in a generalised scheme, involves carefully scheduled production, early morning harvest, immediate fast cooling (e.g. by vacuum cooling), refrigerated transport, trimming, cutting, washing (e.g. with clean, filtered water), disinfecting (e.g. with chlorinated water), drying (e.g. by centrifugation), mixing (e.g. of green and red cabbage), packaging (e.g. vacuum or pillow packs; gas flush with a modified atmosphere), and distribution and sales at low temperature (e.g. 4°C). For fruit, fruit slices (e.g. mango 'cheeks') or pieces (e.g. pomelo segments) are utilised. The modified gas flush atmospheres (low oxygen, high carbon dioxide) minimise oxidative browning, especially at cut surfaces. Important process variables include methods of cutting (e.g. knives, lasers), equipment maintenance (e.g. knife sharpening) and angles of cuts, since such variables determine the degree of tissue wounding. Good hygiene goes hand in hand with low temperature handling, these being critical precautions against the growth of potentially toxic microorganisms (e.g. *Listeria* spp.).

An extension of the lightly processed scenario is fruit drying, which has been long practised for grapes (e.g. sultanas) but which is also applied to many other temperate, subtropical and tropical crops (e.g. mango fruit leather). High temperature and low RH air process conditions minimise browning reactions that might otherwise detract from the appearance of produce.

## Pulsing of cut flowers

Some fresh cut flowers are pulsed, an operation that refers to the supply of a solution via the transpiration stream. In the most basic instance, the chemical is water and the process is hydration. Hydration may be facilitated by the inclusion of a wetting agent in the water. Cut flowers may also be pulsed with sugar, such as sucrose. Sucrose pulsing usually involves concentrated sucrose solutions (e.g. 5–20 per cent w/v) supplied for a matter of hours (e.g. overnight; 16 hours) at room temperature or in the cold room. Sucrose pulsing of gladiolus supplies osmotica and respirable substrate to sustain subsequent bud opening in the vase. Cut flowers such as carnations and delphiniums are pulsed with anti-ethylene agents such as STS or AOA; AOA is toxic to most flowers, except carnations. Typical STS pulsing protocols are 2–4 mM silver ion (as the STS complex) for 15–45 minutes at ambient temperatures or 0.5 mM silver overnight in the cold room at about 1°C. Cut flowers are also pulsed with dyes, such as food grade blue dyes used on white carnations to give interesting visual effects (e.g. blue coloured petal veins and margins).

Cut flowers and foliage destined for desiccation can be pulsed for one to a few days with an humectant, such as 20–30 per cent v/v glycerol. This process is known as uptake preservation. Thus, instead of desiccating completely, plant material treated in this way retains a degree of suppleness (plasticity) associated with the humectant chemical and its attraction of water vapour from the atmosphere into the tissue. As drying plant tissue often browns — particularly when pulsed with an humectant — red, green, blue and other coloured dyes are frequently supplied along with the humectant.

# GRADING

Grading is a critically important process because produce presentation, an aspect of quality, is often judged on the basis of uniformity. Uniformity is important as it presents a standard product for handling (i.e. unitisation) and marketing. Produce is generally graded on the basis of size, weight, colour, defects or composition, or a combination of these features. A typical size grader for fruit separates produce using diverging conveyor belts, which allow smaller fruit to drop between them first. Computerised weight graders can operate on the basis of tipping buckets, which drop to release the preweighed item at a particular position. Video image capture and analysis can also be used for size grading, as well as for colour grading and external defect grading. Video imaging and analysis technology is particularly important for ornamentals, such as cut flowers and pot plants. Composition is not yet widely used as a means of grading fruit. However, automated systems that remove a small tissue core from melons and assess soluble solids concentration (SSC) have been developed and are in commercial use in some parts of the world. Non-destructive near infrared analysis has also proven useful for melons and other fruit that require sorting on the basis of SSC content. Technologies with potential application in grading include X-ray imaging and computer-aided tomography, proton and chemical shift magnetic resonance imaging, transmitted and reflected light spectroscopy, acoustic response signals to tapping, and volatile emissions analysis using solid state detectors (Chapter 10; Table 10.1). However, such advanced technologies are not likely to find practical application until capital cost is minimised and throughput (items/second) is maximised. Nonetheless, given the rapid pace of scientific advances, technological breakthroughs can be anticipated. There is an ever present need to grade out substandard produce on the basis of internal defects (such as jelly seed in mangoes and light brown apple moth in apples) and eating quality (such as acceptable sweetness and flavour in honeydew melon), but these applications are technically much more difficult to achieve.

# FURTHER READING

Akamine, E.K. and J.H. Moy (1983) Delay in postharvest ripening and senescence of fruits, in H.S. Josephson and M.S. Peterson (eds), *Preservation of food by ionising radiation*, vol. 3, CRC Press, Boca Raton, FL, pp. 129–58.

American Society of Heating, Refrigeration and Air-Conditioning Engineers (1986) *ASHRAE Handbook of refrigeration systems and applications*, ASHRAE, Atlanta, GA.

Champ, B.R., E. Highley and G.I. Johnson (1994) *Postharvest handling of tropical fruits*, ACIAR Proc., Australian Centre for International Agricultural Research, Canberra, ACT. no. 50.

Hardenburg, R.E., A.E. Watada and C.Y. Wang (1986) *The commercial storage of fruits, vegetables and florist and nursery stocks* (rev. ed.), US Department of Agriculture, Washington, DC, Agriculture Handbook no. 66.

Hill, A.R., C.J. Rigney and A.N. Sproul (1988) Cold storage of oranges as a disinfestation treatment against fruit flies *Dacus tryoni* (Froggatt) and *Ceratitis capitata* (Wiedemann) (Diptera: Tephritidae), *J. Econ. Entomol.* 81: 257–60.

McCormack, A.A. and W.F. Wardowski (1977) Degreening Florida citrus fruit: Procedures and physiology, *Proc. Internat. Soc. Citriculture* 1: 211–15.

Morris, S.C. (1987) The practical and economic benefits of ionising radiation for the postharvest treatment of fruit and vegetables: An evaluation, *Food Technol. Aust.* 39: 336–41.

Nowak, J. and R.M. Rudnicki (1990) *Postharvest handling and storage of cut flowers, florist greens and potted plants*, Timber Press, Portland, OR.

O'Brien, M., B.F. Cargill and R.B. Fridley (1983) *Principles and practices for harvesting and handling fruits and nuts*, AVI, Westport, CT.

Peleg, K. (1985) *Produce handling, packaging and distribution*, AVI, Westport, CT.

Sawyer, R.L. (1975) Sprout inhibition, in W.F. Talburt and O. Smith (eds), *Potato processing* (3rd ed.), AVI, Westport, CT, pp. 157–69.

Shewfelt, R.L. and S.E. Prussia (eds) (1993) *Postharvest handling — A systems approach*, Academic Press, San Diego, CA.

Story, A. and D.H. Simons (eds) (1997) *Fresh produce manual*, Australian United Fresh Fruit and Vegetable, Brisbane.

World Health Organisation (1977) *Wholesomeness of irradiated food: Report of the joint FAO/IAEA WHO Expert Committee*, WHO, Geneva, series no. 604.

—— (1981) *Wholesomeness of irradiated food: Report of the joint FAO/IAEA WHO Expert Committee*, WHO, Geneva, series no. 659.

# 12
# PACKAGING

The importance of packaging as a determinant of final (consumer level) product quality is largely underrated. Centres of consumption of fresh produce are usually remote from production areas. Accordingly, the allied costs of distribution, which include handling, packaging and transportation, often exceed those of production in both economic and energy terms. Careful management of the distribution system will ensure that produce retains its quality and that economic returns are maximised. Packaging has been practised for as long as fresh produce has been traded. The two main functions of packaging are:

1. to assemble the produce into convenient units for handling (unitisation); and
2. to protect the produce during distribution, storage and marketing (protection).

The earliest packages were mostly constructed of plant materials, such as woven leaves, reeds and grass stems (Figure 12.1), and were designed to be carried by hand (Figure 12.2). Even today, most packages are handled manually at some stage, and are sized accordingly. Nevertheless, packages are often assembled into larger units for mechanical handling by forklift (e.g. pallet loads) or crane (e.g. sea, road and rail containers). Nowadays, produce is transported and sold in an enormous range of packages constructed of wood, fibreboard, jute (hessian) or plastics (Figures 12.3 and 12.4). Many packaging systems waste raw materials, and relatively few packages are reused or recycled, causing waste disposal and environmental problems.

Modern packages and packaging for fresh produce is expected to meet a range of basic requirements. They must:

- have sufficient mechanical strength to protect the contents during handling and transport, and while stacked;
- be largely unaffected, in terms of mechanical strength, by moisture content when wet or at high RH;
- stabilise and secure the product against movement within the package during handling;
- not contain chemicals that could become transferred to the produce and taint it or be toxic to the produce or to humans;

- meet handling and marketing requirements in terms of weight, size and shape;
- allow rapid cooling of the contents and/or offer a degree of insulation from external heat or cold;
- use gas barriers (e.g. plastic films) with sufficient permeability to respiratory gases in order to avoid any risk of anaerobiosis;
- offer security for the contents and/or ease of opening and closing in some marketing situations;
- identify the contents, proffer handling instructions and aid retail presentation through comprehensive and accurate labelling;
- either exclude light (e.g. from potatoes) or be transparent (e.g. for orchids);
- facilitate easy disposal, reuse or recycling; and
- be cost-effective in relation to the value and the required extent of protection of the contents.

It is becoming increasingly important to reduce the many types (sizes and shapes) of packages through standardisation. Unitisation (e.g. pallets) and mechanical handling (e.g. forklifts) make standardisation essential for economical operation.

**Figure 12.1**
Typical bamboo basket used throughout South-East Asia for the handling and transportation of produce.

**Figure 12.2**
Shoulder transport of taro in bamboo baskets in Indonesia.

# PREVENTION OF MECHANICAL DAMAGE

Fruit and vegetables vary widely in their susceptibility to mechanical damage and in the types of mechanical injury to which they are susceptible. Accordingly, the choice of package and packing method must take account of these differences. Four different causes of mechanical injury to produce can be identified: vibration rubs, compressions, impacts and cuts. The relative susceptibilities of some fruits to compression, impact and vibration injuries are shown in Table 12.1.

With the possible exception of certain so-called 'hard' vegetables such as watermelons, pumpkins, onions, carrots and potatoes, the package must be strong enough to carry stacking loads or there will be compression bruising of the contents. Impact bruising is caused by dropping packages and by other impact shocks during handling (e.g. on the grading line) and transport. Bruising is tissue damage that results from strain energy being dissipated in the tissue. The amount of damage depends on how much energy is dissipated and the nature of the tissue. Vibration injury is common during transport, resulting in abrasion marks ranging from light rubbing to removal of the skin and possibly some of the flesh. The problem of cuts (e.g. sharp edges) and punctures (e.g. nails, stems) also merits attention.

Nearly all mechanically damaged tissue turns brown through enzymatic (e.g. polyphenol oxidase) or chemical (atmospheric oxygen) oxidation of phenols (producing tannins). Tissue deformation and browning is exacerbated by increased water loss as a result of damage to the cuticle. Furthermore, injuries represent sites for microbial infection in terms of both damage to the protective skin and release of substrates for growth (e.g. sugars). The physiological wound response involves increased ethylene production and respiration, and hence acceleration in the rate of produce deterioration. Mechanical damage also causes an immediate loss of edible material because of the need to trim off unsightly portions or to discard whole units.

Two important practical requirements must be met when packaging perishable produce:

◆ individual items should not be allowed to move with respect to each other or to the walls of the package in order to avoid vibration injury; and

◆ the package should be full without packing too tightly, which increases compression and impact bruising.

In short, both underfilling and overfilling are to be avoided so that individual items are held firmly, but not too tightly, within the package.

**Figure 12.3**
(top) Bulk bins made from fibreboard. (bottom) Produce in fibreboard cartons.

Packaging can be made more protective by individually wrapping each individual product (e.g. paper wraps), by isolating each piece (e.g. cell and tray packs) or by use of energy-absorbing materials (e.g. cushioning pads). However, such additional packaging increases costs, and must therefore be justified by reduced waste, increased selling price, or the establishment of a reputation for reliable quality so that price and sales are maintained during periods of oversupply. Careful handling of the packages is the best general precaution against mechanical damage.

The potential for vibration damage during transport deserves particular emphasis. Each product will start to vibrate (move around or

**217**

**Figure 12.4**
(top) Boxes made of foamed polystyrene. (bottom) (back) onions in plastic mesh bags; (front left) carrots in large plastic bags; (front centre and right) potatoes in woven jute and plastic bags.

rotate) at a certain excitation (resonant) frequency (Hz) that is specific to each product. In the case of produce packaged in stacks (e.g. palletised) during road transport, the vibrations resulting from the interaction of the road surface and the vehicle suspension system (2–20 Hz) will also be propagated with increasing magnitude up through the stack. It is for these reasons that displaced cartons and produce affected by vibration injury are often seen at the top of stacks.

## TABLE 12.1  SUSCEPTIBILITY OF PRODUCE TO TYPES OF MECHANICAL INJURY

| PRODUCE | TYPE OF INJURY | | |
| --- | --- | --- | --- |
| | COMPRESSION | IMPACT | VIBRATION |
| Apple | S | S | I |
| Apricot | I | I | S |
| Banana, green | I | I | S |
| Banana, ripe | S | S | S |
| Cantaloupe | S | I | I |
| Grape | R | I | S |
| Nectarine | I | I | S |
| Peach | S | S | S |
| Pear | R | I | S |
| Plum | R | R | S |
| Strawberry | S | I | R |
| Summer squash | I | S | S |
| Tomato, green | S | I | I |
| Tomato, pink | S | S | I |

S — susceptible;
I — intermediate;
R — resistant.

SOURCE  R. Guillou (1964) *Orderly development of produce containers, Proceeding, Fruit and Vegetable Perishables Handling Conference, 23–25 March 1964*, University of California, Davis, CA. pp. 20–25.

# COOLING PRODUCE IN THE PACKAGE

A further important requirement of packaging is that it may be necessary to allow rapid cooling of produce. For example, containers designed for pressure cooling (Chapter 4) should have holes that occupy about 5 per cent of the surface area on each of the air entry and exit ends. Both the nature of the produce and the treatment after packing must be considered in designing containers for specific commodities. Ideally, respiratory heat should be able to escape readily from the package. In the case of small and/or tightly packed commodities, such as green beans, peas, small fruits, leafy vegetables and cut flowers and foliage, the vital heat of respiration is removed largely by conduction to the surface of the package. Thus, the volume of the contents (i.e. the minimum dimension of the package from the centre to the surface) becomes a critical issue. The acceptable volume depends on the respiration rate of the commodity. If the volume of the package filled with produce is excessive, produce near the centre of the package will heat up because respiratory heat cannot dissipate fast enough. In practice, self-heating is a significant problem for rapidly respiring produce such as peas, beans, lettuce and broccoli in large single packages or in close stacks of packages. The problem may be avoided with small-er packages or by ventilating large packages or stacks to allow air circulation.

Bulk bins that hold about 500 kg of produce are now widely used for harvesting, storage and often transport of fruit and vegetables. Adequate cooling of fruit in stacks of bins may be achieved in cool stores without forced movement of air, through a 'chimney effect', if about 10 per cent of the floor area of these bins is vented. Nevertheless, fan-assisted ventilation is necessary for rapid overnight cooling

# EFFECT OF PACKAGING ON WEIGHT LOSS

Packaging is often used to minimise weight loss and shrinkage of produce during marketing. Specific approaches include:

◆ waxed paper or plastic film wrapping or bagging of individual items;
◆ consumer packaging of a number of items in trays (e.g. polystyrene) overwrapped with a moisture barrier material; and
◆ increasing the moisture resistance of fibreboard packages with a surface or internal plastic laminate or by lightly or fully waxing the carton.

Under dry conditions, containers of produce (e.g. wooden boxes, plastic crates, baskets) may be deliberately sprayed with water. Direct wetting can also assist in cooling produce by evaporation of the added water. Fresh cut flowers and foliage are often transported 'wet', usually in plastic buckets but sometimes with stems in individual phials of solution.

# PACKAGE DIMENSIONS

Package dimensions are both economically and structurally important. The size and shape should facilitate economic use of materials. Packages should also offer adequate strength and allow easy and secure handling, loading and stacking. An optimal length to width ratio is about 1.5:1. There is a trend towards the use of smaller packages because of recommendations by the International Labour Organization concerning maximum weights that people can reasonably be expected to handle routinely. Thirty litre (about 20 kg of produce) and 15 L packages are becoming standard for fruit, with larger 36 L packages being used for some vegetables.

Standardisation of package sizes promotes efficient handling. Package sizes are standardised by determining dimensions and numbers that fit well on standard pallets (Figure 12.5). However, achieving this objective may require that the 'produce be grown to fit the package' rather than the 'package be sized to fit the produce', as in the past. Traditional baskets and hessian or plastic mesh bags, although relatively cheap, do not stack efficiently. Also, they are often too large for efficient handling and there is a tendency to overpack them with produce (Figure 12.1, page 215). Produce therein is often damaged during handling and transport because of a lack of structural support to protect the contents.

**Figure 12.5** Handling pallet loads of packaged produce in a room used for temporary cool storage. Note the mixture of boxes of different sizes in front of the cooling unit.

# MECHANICAL STRENGTH OF THE PACKAGE
·······

For continued protection of produce against mechanical damage, packages must retain their strength throughout the marketing chain. A common saying is that: 'the package should support the produce, not the produce should support the package'. Wood, solid plastic and expanded plastic packages are inherently strong compared with fibreboard packages. However, wood is an increasingly expensive and environmentally costly material. Solid plastic containers are even more expensive, although they are amenable to washing and reuse. Rigid expanded polystyrene is lightweight and yet strong, but requires considerable storage space and must be recycled using expensive capital infrastructure into high density material. In comparison, fibreboard is attractive and can be made stronger by using two or more thicknesses, such as the bottom and lid of fully telescoping cartons. The strength of fibreboard lies in the fluting between the inner and outer liners. Fibreboard comprised of two layers of fluting sandwiched between three liner layers is stronger than the conventional single layer of fluting.

Under conditions of high humidity, after condensation or after being wet by rain, the strength of the package must either be independent of moisture content (e.g. wood) or the package material must not absorb moisture (e.g. plastic). Commonly used fibreboard cartons and trays rapidly lose strength as they absorb moisture and therefore are less than satisfactory under tropical conditions and in high humidity cool storage. However, fibreboard can be protected if fully impregnated with wax or similar material, although wax impregnation is expensive and waxed fibreboard is not recyclable. The corrugated fibreboard used in cartons is often only given a surface coating (termed 'lightly waxed') that affords only a degree of protection from free water or high humidity. Packages and/or packing materials might also be required to exclude water from the produce (e.g. to prevent grape splitting) or to prevent dehydration of the produce (e.g. roses that otherwise suffer bent neck).

# PACKAGE TESTING
·······

Testing the performance of packages is an essential part of package development. However, the impacts and vibrations that occur during commercial handling and transport of packaged produce are variable and complex, and therefore difficult to simulate in the laboratory. Useful laboratory tests include machine testing the compressive strength of packaging under standard conditions, and simple drop tests. Quite complex transport shock simulators, which are programmed from actual transport operation records using strain gauges and other electronic sensors, have been developed. Such simulators are expensive and can only be justified for specialised research. There is no complete substitute for monitoring package performance under actual commercial conditions.

## PACKING

The desirable pack for most fruit and vegetables is one in which the package is tightly filled; that is, without underfilling or overfilling. The package, and not the produce, should bear the stacking load. However, it is equally important that produce does not move and sustain vibration injury during handling and transport. Some produce, such as potatoes, onions and carrots, and some types of citrus fruit (e.g. oranges), will withstand reasonable compressive loads. These can be satisfactorily packaged in non-rigid packages, such as mesh bags, provided they are handled with due care. In a generally similar fashion, some vegetables (e.g. asparagus) and cut flowers (e.g. irises, gladioli) can be packaged in bundles.

In many countries, fruit has traditionally been place- or pattern-packed, such that each piece was put into a specific position by hand. The objectives of place packing were to maximise net weight, maintain a tight pack, and present the fruit attractively when the package was opened. However, this time-consuming approach, which requires accurate grading into a range of size-classes, has become very costly. The alternative is volume- or tight-fill packing, whereby fruit are 'poured' into the carton or box. After filling, the pack is vibrated to obtain a reasonably tight pack of the fruit within. By this method, produce (e.g. apples) is packed to a standard weight rather than a standard count. In a volume-fill pack, produce is often kept firmly positioned beneath a pressure pad (e.g. paper pulp in an envelope) over which the lid is securely fastened. Package inserts that isolate individual fruits, such as moulded pulp or plastic trays, are expensive. Nonetheless, they remain in common use for delicate and/or high value products (e.g. mangoes). Individual wrapping still has a place, as does lining the package with plastic or paper in order to reduce vibration damage and/or moisture loss. Very delicate fruit, such as papaya (pawpaw), may be sleeved in thick spongy plastic mesh. Cut flowers are often packaged in paper or wood 'wool' for protection, and flower bunches and potted plants are often sleeved in plastic film.

It is not practical to recommend specific packages for each fruit, vegetable or ornamental, as several types may be satisfactory. The most suitable package would depend on many factors, including the region, environmental conditions, length and nature of the market chain, methods of handling and transport, availability and cost of materials, and whether the produce is to be refrigerated.

# STOWAGE

Stowage or stacking of packages should simultaneously ensure stability of the stack and allow adequate air movement to enable satisfactory cooling. Stowage must also be economical of space and easy to achieve. Stronger packages are needed for long distance transport and/or for high stacking in cool stores. Where packages are normally handled on pallets, the package must fit the pallet. Stack stability is usually best obtained by either cross-stacking or tied-stacking, which require the packages to have suitable length to breadth ratios. Tied-stacking can be achieved by straps, nets, plastic wrap, tape and glue.

# CONSUMER PREPACKING

Produce purchased at retail outlets was traditionally packaged in paper bags. However, paper bags have been largely replaced by polyethylene film bags, which are cheaper, stronger and usually transparent. Supermarket outlets have led a trend towards marketing prepackaged fruit and vegetables, whereby produce is preweighed and packaged into small units for retail sale. However, consumers are sometimes wary of the quality of prepackaged items, and many still opt to select individual produce items from an open display.

Consumer packaging of produce in small plastic bags or in plastic or paperboard trays overwrapped with clear film (Figure 12.6) helps restrict weight loss from produce. Such packaging can also provide a modified atmosphere benefit. However, the modified atmosphere effect is somewhat risky and generally not sought. Accordingly, films with low gas permeability are perforated to prevent significant modification of the package atmosphere. The number of small perforations necessary does not increase water loss appreciably.

Plastic films for packaging produce have good tensile strength under all likely environmental conditions. Other properties — such as gas and water permeability, heat sealability, clarity and printability — can generally be specified by the packer. Manufacturers can develop a film to most specifications. Low density polyethylene film is most widely used for consumer packs. Polyethylene has good clarity, can be heat sealed, is flexible over a wide temperature range (–50–70°C), and is probably the cheapest film in most countries. Polyethylene is relatively permeable to many volatile compounds and gases, but is comparatively impermeable to water vapour. Gas permeability of films can be controlled by varying either the density of the film or its thickness, or, as mentioned above, the film may be perforated.

**Figure 12.6**
(top) Self-service displays of prepackaged and loose fruit and vegetables in a modern supermarket.
(bottom) Prepackaged whole and fresh-cut vegetables in a refrigerated display cabinet.
(Reproduced with permission of Woolworths P/L, Australia.)

# MODERN PACKAGING
·······

With the rapid advance of technology, there is interest in producing increasingly sophisticated packaging. The term 'active packaging' has been coined to describe packaging that offers a level of control over in-package conditions and how they vary with product (e.g. ethylene production) and environmental (e.g. temperature) factors. An example might be a polymer film whereby the permeability to oxygen and carbon dioxide can increase or decrease as temperatures rise and fall, respectively. In the context of modern materials science, it is only a matter of time before such films are commercially available.

Water loss in packages can be reduced by the use of entire or micro-perforated (pin-hole size) or macro-perforated films. However, condensation within moisture barriers can become a problem. In film-wrapped produce (e.g. citrus wrapped in heat shrink film), condensation is generally not a problem because the film is in intimate contact with the fruit and assumes the same temperature as the fruit. In the case of loosely wrapped produce (e.g. cut flowers within a plastic carton liner or fruit in a consumer pack), condensation can be reduced with simple or elaborate moisture sinks, such as newspaper and salts in spun-bonded polyethylene, respectively. Chemical anti-fogging treatments can be applied to films, and films with relatively high water permeability can be used (e.g. cellophane, polyvinyl chloride). Importantly, these precautions to avoid condensation also maintain product visibility (e.g. blueberries in overwrapped punnets). Water absorbents can be incorporated in packaging to capture and hold the free water resulting initially from condensation, which is followed by droplet formation and finally pooling.

Ethylene can be scrubbed from the package using blocks or sachets of high surface area materials (e.g. florists' foam, aluminium oxide particles) coated with potassium permanganate or using films impregnated with another oxidant, such as tetrazine. Tetrazine is a particularly promising compound since, unlike potassium permanganate, it is relatively specific for small double-bonded volatiles like ethylene. However, whatever the active ingredient, ethylene scrubbers are generally unlikely to work effectively unless they are positioned solely to intercept incoming ethylene. Ethylene is deleterious at ppb levels and is produced by the harvested plant tissue itself. Concentration differences at ppb levels do not constitute gradients of sufficient magnitude to drive ethylene to a point (e.g. sachet) or planar (e.g. film) sink. Thus the provision of ethylene scrubbers within packages can sometimes be a waste of money and effort. It is tempting to envisage small ethylene scrubbers with tiny motorised fans that draw in the in-package air in order to lower in-package ethylene concentrations.

Respiratory gas levels (i.e. oxygen, carbon dioxide) can be controlled by chemical (e.g. polymer type) and physical (e.g. thickness) characteristics of plastic films, as well as by holes in films. Oxygen and carbon dioxide flux through holes is proportionally greater in magnitude than water vapour and ethylene flux because their flux is driven by comparatively large concentration gradients (per cent versus ppm or ppb). Thus, perforated films can be used effectively to reduce water loss while avoiding the risk of anaerobiosis. All other factors being equal, oxygen diffuses somewhat faster in air (e.g. through holes) than carbon dioxide on account of its greater diffusion coefficient. In contrast, all plastic films are relatively more permeable to carbon dioxide than oxygen. Both oxygen and carbon dioxide can be chemically scrubbed from packages. Lack of consistent temperature control during handling and transport and differences in the temperature quotient for physical gas diffusion across plastic films (smaller $Q_{10}$) compared with those for physiological processes such as respiration (larger $Q_{10}$) increase the likelihood that anaerobic conditions may occur in sealed plastic film packages. Such risks may be minimised using failsafe (e.g. low melting point polymers) or variable aperture (e.g. bimetallic strips) devices to regulate the formation and/or size of holes. As suggested earlier, advances in microelectronic, biosensor and polymer sciences are likely to give films that actively sense and respond in a controlled way to stimuli, such as an increase in temperature.

In the absence of refrigeration, a certain level of temperature control during handling and transport can be achieved with insulation (e.g. polystyrene boxes) and with heat sinks provided inside the packaging (e.g. loose ice, ice packs and gel ice packs). In the absence of refrigeration, external reflective and/or insulative covers (e.g. thermal blankets) and heat sinks (e.g. dry ice) can serve to assist or provide an alternative to in-package temperature control measures.

# ENVIRONMENTAL ISSUES

Packaging for horticulture produce has enormous environmental implications because of the large quantities of material involved and their eventual disuse. Wooden packaging is biodegradable, as is fibreboard and paper packaging, which is also recyclable. However, these materials are forest products, and forest resources have not been and are not always going to be managed responsibly. Solid plastic packaging is mostly reusable. Unfortunately, however, returnable plastic crates are relatively expensive, both in terms of initial outlay (production, purchase or hire) and maintenance costs (e.g. washing, backloading). Polystyrene (expanded polymer) and plastic films constitute significant environmental problems, both aesthetically and to wildlife. Nevertheless, polystyrene can and is being recycled (e.g. to high density material). Plastic films can also be recycled, although stringent recycling practices are seldom applied at present. Nevertheless, there is little question that, through public pressure, more environmentally sound packaging will come into use. For example, biodegradable packaging films made from materials such as cellulose, starch and proteins are being developed. In time, if not at present, the use of environmentally sound packaging should offer a marketing edge.

# FURTHER READING

......

American Society of Heating, Refrigeration and Air-Conditioning Engineers (1986) *ASHRAE Handbook of Refrigeration Systems and Applications*, ASHRAE, Atlanta, GA.

Champ, B.R., E. Highley and G.I. Johnson (1994) Postharvest handling of tropical fruits, *ACIAR Proc.*, Australian Centre for International Agricultural Research, Canberra, ACT. no. 50.

Hardenburg, R.E., A.E. Watada and C.Y. Wang (1986) *The commercial storage of fruits, vegetables and florist and nursery stocks* (rev. ed.), US Department of Agriculture, Washington, DC. Agriculture Handbook no. 66.

Holt, J.E. and D. Schoorl (1984) Package protection and energy dissipation in apple packs, *Sci. Hort.* 24: 165–76.

Nowak, J. and R.M. Rudnicki (1990) *Postharvest handling and storage of cut flowers, florist greens and potted plants*, Timber Press, Portland, OR.

O'Brien, M., B.F. Cargill and R.B. Fridley (1983) *Principles and practices for harvesting and handling fruits and nuts*, AVI, Westport, CT.

Peleg, K. (1985) *Produce handling, packaging and distribution*, AVI, Westport, CT.

Shewfelt, R.L. and S.E. Prussia (eds) (1993) *Postharvest handling — A systems approach*, Academic Press, San Diego, CA.

Story, A. and D.H. Simons (eds) (1997) *Fresh produce manual*, Australian United Fresh Fruit and Vegetable Association Ltd, Sydney.

# 13
# COMMODITY STORAGE RECOMMENDATIONS

The aim of postharvest technology is to provide producers and traders with flexibility in when and where to market commodities in order to obtain the maximum net cash return. Earlier chapters have indicated a range of technologies available to extend postharvest life. These technologies range from simple operational management such as careful manual handling of produce to minimise mechanical damage to sophisticated electronic systems in regulated CA stores. Technologies available for some produce such as certain apple cultivars will allow a storage life of 12 months, whereas the extension in market life of produce such as highly perishable berry fruit and leafy vegetables may only be a few days. All postharvest interventions, whether from changes in handling, storage or transport systems, come at some cost that must be recouped from an enhanced market return. It is therefore a subjective value judgment as to when and where to market and therefore what postharvest technology is required.

The root cause of postharvest deterioration that needs to be inhibited is enhanced metabolism — whether due to natural senescence physiology or biotic (e.g. pathogen) or abiotic (e.g. physical injury) stress — with the main technological interventions being to control the temperature and humidity of the atmosphere around produce. For an ideal commodity, the optimum temperature is just above the freezing point and optimum relative humidity is 100 per cent. However, as discussed previously, there are few ideal commodities. Tor example, the onset of low temperature injuries and microbial growth will reduce storage life. Recommended storage conditions therefore become a compromise to obtain the maximum time without any adverse reaction to the imposed environmental conditions, and at a financial cost.

When consulting published tables of recommended storage conditions, it needs to be remembered that such tables have been compiled from data generated by many research groups in many countries. They therefore should only be accepted as a guide to likely storage behaviour for any particular commodity. The precise optimal storage conditions for a specific variety of a specific commodity grown in a specific locality needs to be determined experimentally.

## *Temperature*

Table 13.1 gives recommended temperature ranges for a selection of fruit and vegetables. The listing under temperature ranges recognises that, within a particular produce type, factors such as cultivar, season and maturity at harvest can give different physiological and metabolic responses that result in variations in the optimum storage temperature. For example, the optimum storage temperature reported for apples ranges from —1°C to 5°C. If the volume being handled is insufficient to fill a storage room or transport container, a number of different produce may be held in the same storage chamber and therefore a temperature compromise is required. A meaningful temperature setting for the long-term mixed storage of horticultural produce could be 0°C for produce not susceptible to low temperature injuries, 5°C for produce susceptible to physiological disorders and 10°C for produce susceptible to chilling injury. For produce that are highly susceptible to chilling injury or where further maturation is required, ambient storage is recommended (see Chapter 4). Considerations of compatibility of humidity requirements (see Chapter 5) and ethylene sensitivity (see Chapter 6) will also determine the feasibility of mixed storage arrangements.

It must be remembered that a time/temperature relationship exists for the development of low temperature injuries so that temperature-sensitive produce can be safely held at suboptimal temperatures for short periods (see Chapter 8). This means that where short-term mixed storage is required for up to about 1 week, such as in transit to markets or holding in wholesale or retail markets, a more liberal temperature regime can apply. A large number of produce can be safely stored together at 0°C with most others held at 7–10°C. Some tropical produce, potatoes, onions and pumpkins can be held at 15°C or even at ambient temperature.

The storage times presented in Table 13.1 show a wide variation, ranging from a few days to many months. While the storage life of individual produce is governed by many factors, an overriding determinant as to whether it has a short or long storage life is the overall rate of metabolism, which is related to the respiration rate. Produce with a low respiration rate generally store longer. Some examples of produce respiration rates are given in Table 13.2. Those items with the shortest

## TABLE 13.1  RECOMMENDED TEMPERATURE TO MAXIMISE STORAGE LIFE OF SELECTED FRUIT AND VEGETABLES

| PRODUCE | TIME AT OPTIMUM TEMPERATURE (WEEKS) | | |
|---|---|---|---|
| | 1–4°C | 5–9°C | 10°C+ |
| **Fruit** | | | |
| *Very perishable (0–4 weeks)* | | | |
| Apricot | 2 | | |
| Banana, green | | | 1–2 |
| Berry fruits | 1–2 | | |
| Cherry | 1–4 | | |
| Fig | 2–3 | | |
| Mango | | | 2–3 |
| Strawberry | 1–5 days | | |
| Watermelon | | 2–3 | |
| *Perishable (4–8 weeks)* | | | |
| Avocado | | 3–5 | |
| Grape | 4–6 | | |
| Mandarin | | 4–6 | |
| Passionfruit | | 4–5 | |
| Peach | 2–6 | | |
| Pineapple, green | | | 4–5 |
| Plum | 2–7 | | |
| *Semi-perishable (6–12 weeks)* | | | |
| Coconut | 8–12 | | |
| Orange | | 6–12 | |
| *Non-perishable (> 12 weeks)* | | | |
| Apple | 8–30 | | |
| Grapefruit | | | 12–16 |
| Pear | 8–30 | | |
| **Vegetables** (incl. fruit vegetables) | | | |
| *Very perishable (0–4 weeks)* | | | |
| Asparagus | 2–4 | | |
| Bean | 1–3 | | |
| Broccoli | 1–2 | | |
| Cucumber | | 2–4 | |
| Lettuce | 1–3 | | |
| Mushroom | 2–3 | | |
| Pea | 1–3 | | |
| Tomato | | | 1–3 |
| *Perishable (4–8 weeks)* | | | |
| Cabbage | 4–8 | | |
| *Semi-perishable (6–12 weeks)* | | | |
| Celery | 6–10 | | |
| Leek | 8–12 | | |
| Marrow | | | 6–10 |
| *Non-perishable (> 12 weeks)* | | | |
| Carrot | 12–20 | | |
| Onion | 12–28 | | |
| Potato | | 16–24 | |
| Pumpkin | | | 12–24 |
| Sweet potato | | | 16–24 |

TABLE 13.2 TYPICAL RESPIRATION RATES OF
SELECTED FRUIT AND VEGETABLES

| FRUIT | RESPIRATION RATE AT 15°C (ML $CO_2$/KG/H) | VEGETABLE | RESPIRATION RATE AT 15°C (ML $CO_2$/KG/H) |
|---|---|---|---|
| Apple | 25 | Bean | 250 |
| Banana, green | 45 | Cabbage | 32 |
| Banana, ripe | 200 | Carrot | 45 |
| Grape | 16 | Lettuce | 200 |
| Lemon | 20 | Pea | 260 |
| Orange | 20 | Potato | 8 |
| Peach | 50 | | |
| Pear | 70 | | |
| Strawberry | 75 | | |

SOURCE Adapted from American Society of Heating, Refrigeration and Air-conditioning Engineers (1986) *ASHRAE handbook of refrigeration systems and applications*, ASHRAE, Atlanta, GA.

storage life are thus leafy vegetables and fruits, such as berries that are harvested when ripe, that have a high respiration rate, and chilling-sensitive tropical fruit that cannot be held at low temperature to decrease metabolism. Produce with the longest storage life are underground vegetables and some pome and citrus fruit that have a relatively low respiration rate and/or can tolerate low temperature storage.

The termination of storage life in many produce, even at low temperatures, is at the onset of microbial growth. The ability of produce to withstand microbial invasion is related not only to external environmental conditions but also to the structural integrity of outer cells. The onset of microbial growth is thus often due to natural metabolism during storage, leading to general senescence which promotes spore germination and mycelial growth of existing quiescent infections rather than any new microbial infection of produce (see Chapter 9).

### *Humidity*

A high relative humidity around produce is required to minimise water loss. As discussed in Chapter 5, the use of 100 per cent RH can create problems due to condensation of water leading to the development of rotting. A saturated atmosphere can only be utilised where produce have some resistance to rotting or are prone to excessive rates of water loss. Recommended humidity conditions then become a compromise between reducing water loss and preventing microbial growth, with the latter factor being paramount. For most produce, recommendations tend to be in the range of 85–95 per cent RH, but can be 98 per cent for produce with very high transpiration rates and 60 per cent for produce highly susceptible to rotting.

# GROUP COMMODITY RECOMMENDATIONS
.......

## Leafy vegetables and immature flower heads
These produce have high transpiration rates due to the high ratio of surface area to volume and many have high rates of respiration. They should be cooled to about 0°C as soon as practical after harvest and maintained at low temperature throughout storage and preferably also during marketing. The RH should be at least 95 per cent. It has been suggested that 100 per cent RH can be sustained provided the temperature is maintained not higher than 0°C. Some consideration needs to be given to the removal of the heat of respiration from the more immature vegetables; this may require ventilation with high humidity air.

## Vegetable fruits
Vegetable fruits can be divided into two subgroups: those consumed in an immature unripe condition (such as green legumes, cucumbers, eggplant and peppers) and those consumed when mature and ripe (such as melons, pumpkins and tomatoes). Apart from green peas, which can be stored at 0°C, most vegetable fruits are susceptible to low temperature injuries to some extent and are therefore stored at temperatures ranging from 3–5°C for beans to 10–15°C for pumpkins. Tomatoes, in common with many other climacteric fruits, are more tolerant to low temperatures when ripe. A high relative humidity (>95 per cent) is essential for peas and beans but a low humidity (about 60 per cent) is required for pumpkins and winter squash, with most other fruit vegetables able to be held at 90–95 per cent.

## Underground vegetables
This group encompasses diverse produce types covering bulbs, roots and

tubers and other plant material such as the rhizomes (e.g. ginger). Most are characterised by a low respiration rate and hence tend to have a relatively long storage life. Bulbs can be stored at 0°C or slightly lower but need to have a low RH at about 70 per cent to prevent rotting and root growth. Long storage bulbs should be cured before storage. They should not be held for any period at 5–20°C as rotting and sprouting occurs most rapidly at these temperatures. Most of the temperate root vegetables are recommended for storage at 0°C and at least 95 per cent RH. Sweet potato, which is a fleshy storage root, is susceptible to chilling injury at temperatures below 10°C; it benefits from curing before storage. Most tubers are susceptible to low temperature injuries and should be held at temperatures ranging from 5∞C–15°C. Sprouting and rotting are the major storage problems and most root vegetables benefit from curing (see Chapter 11).

## Deciduous tree and vine fruit

The storage life of deciduous fruit is variable, ranging from 1–2 weeks for apricot and figs to more than 6 months for some apples and pears (Table 13.1). Species with a short storage life benefit from rapid cooling after harvest. The recommended storage temperature for most deciduous fruit is −1°C–0°C, except for specific apple and pear cultivars that are susceptible to physiological disorders when the recommended temperature can be up to 5°C. While deciduous fruit with a relatively short storage life are also generally susceptible to low temperature injuries, 0°C remains the recommended temperature as an even shorter storage life through enhanced senescence occurs at even slightly higher temperatures. Care must be taken with deciduous fruit stored in an unripe condition as prolonged storage at low temperature can inhibit the ability of fruit to ripen after cool storage. European cultivars of pears are particularly susceptible to this effect. The RH during storage should be 90–95 per cent.

## Berries

As a group, berries are probably the most perishable of all fruit and vegetables, with some such as blackberries having only a few days storage life even under optimal conditions. Berries are non-climacteric fruit with high respiration rates, and are harvested at optimum eating quality. Their soft texture makes them highly susceptible to physical damage, leading to general senescence and rotting. The recommended storage temperature for most berries is −1°C–0°C at 90–95 per cent RH. Berries benefit from rapid cooling soon after harvest.

## Citrus fruit

All citrus fruit are prone to developing a wide range of physiological disorders, but susceptibility varies greatly between citrus species and between cultivars and often growing region within species (see Chapter 8). This gives rise to a wide range of recommended storage temperatures. General recommendations range from 4°C for mandarins to 15°C for grapefruit. Rotting and water loss can be problems during prolonged storage, and an RH of 85–90 per cent is recommended, although waxing with added fungicides is often practiced to overcome these problems (see Chapter 9).

## Tropical and subtropical fruit

Most tropical and subtropical fruit are susceptible to low temperature injuries, with the level of severity often related to the background temperature of the normal growing environment (see Chapter 8). Most tropical fruit are highly susceptible to chilling injury and should not be stored below 10°C, whereas some subtropical fruit such as avocados have a wide range of recommended storage temperatures from 4–13°C and kiwifruit can be stored at 0°C. Ripe fruit can often be stored at temperatures about 5°C lower than unripe fruit, albeit with a limited storage life. The recommended RH is in the range 85–95 per cent.

## Ornamentals

Fresh cut flowers and foliage typically comprise immature organs (e.g. flower buds and leaves) and often have high surface area to volume ratios. Thus, they often have high metabolic rates and high rates of water loss. Accordingly, they should be stored at the lowest temperature they can tolerate and at high RH. Generic storage recommendations for such materials are typically 0.5°C and ≥95 per cent RH. Such conditions are generally applicable to the major commercial cut flower crops but a number of cut flower and foliage lines, particularly of subtropical and tropical origin, are chilling-sensitive. Chilling-sensitive species should be held at temperatures of 13°C or above, the precise temperature depending on species, cultivar, stage of maturity, production environment and duration of storage. Cut flowers and foliage can be stored wet (i.e. standing in solution) or dry. For longer-term storage, the dry option is preferred because it is more efficient with space, the plant's metabolism is lower, and the potential for decay is reduced. Cut flowers may be given various treatments prior to storage (e.g. pulsed with sucrose solution) to help maintain quality. Examples of recommended storage conditions for a range of cut flowers are given in Table 13.3.

Potted plants are generally not stored for significant periods of time.

## TABLE 13.3  STORAGE CONDITIONS FOR 20 SELECTED CUT FLOWERS

| Cut flower | Temp (°C) | RH (%) | Storage life (days) | Short-term storage temp (°C) |
|---|---|---|---|---|
| Alstroemeria* | 0–4 | 90–95 | 6–10 | 1 |
| Anthurium* | 12.5–15.5 | 90–95 | 3–10 | 15 |
| Bird-of-paradise | 7–10 | 85–95 | 3–28 | 7.5 |
| Carnation* | 0–7 | 90–95 | 3–42 | 1 |
| Chrysanthemum | -0.5–8 | 90–98 | 7–42 | 1 |
| Delphinium* | 0–4.5 | 90–95 | 1–2 | – |
| Freesia* | 0–4 | 90–95 | 1–14 | 1 |
| Ginger | 7–10 | 90–95 | 5 | – |
| Gypsophila* | 0–4.5 | 98 | 1–21 | 1 |
| Iris§ | -0.5–4 | 90–95 | 4–28 | 1 |
| Liatris | 0–5 | 90–95 | 3–14 | – |
| Lily* | 0–4.5 | 90–95 | 4–28 | 1 |
| Lisianthus | 1 | 90–95 | 7 | 1 |
| Narcissus§ | 0–2 | 90–95 | 7–21 | 1 |
| Orchid† | 0.5–15 | 90–95 | 7–28 | – |
| Protea | 2–4 | – | 21 | 2 |
| Rose* | 0–4 | 90–98 | 4–14 | 1 |
| Snapdragon* | -1–5 | – | 3–28 | 1 |
| Statice | 1.5–4 | 90–95 | 14–42 | 2 |
| Tulip§ | -0.5–2 | 85–95 | 3–42 | 1 |

† Tropical orchids (e.g. *Vanda*, *Dendrobium*, *Cattleya*) may be stored at 5–15°C, whereas *Cymbidium* and *Paphiopedilum* may be stored at lower temperatures (e.g.-0.5–4°C).
* Cut flowers for which STS pulsing has been recommended.
§ Cut flowers reported to be sensitive to ethylene, but for which specific STS recommendations have not been made.

SOURCES  Principal reference: R.E. Hardenburg, A.E. Watada and C.Y. Wang (1990) *The commercial storage of fruits, vegetables and florist and nursery stocks* (rev. ed.), US Department of Agriculture, Washington, DC. Handbook no. 66. Also: R. Jones (1991) *Post-harvest care of cut flowers*, Institute of Plant Sciences, Department of Agriculture, Knoxfield, Victoria. 72 pp.; D. Joyce (1988) *Storage conditions for ornamental crops*, Western Australian Department of Agriculture, South Perth. Farmnote no. 34/88; J. Nowak and R.M. Rudnicki (1990) *Postharvest handling and storage of cut flowers, florist greens and potted plants*, Timber Press, Portland, OR.

## TABLE 13.4  RECOMMENDED STORAGE TEMPERATURE FOR 10 SELECTED CUT FOLIAGE AND POT PLANTS

| PLANT | (°C) |
|---|---|
| A. *Cut foliage** | |
| Asparagus | 0–5 |
| Dieffenbachia | 13 |
| Eucalyptus | 1.5–5 |
| Holly | 0–2 |
| Maidenhair fern | 0–4.5 |
| B. *Pot plants*† | |
| Aglaonema | 16–18 |
| Begonia | 10 |
| Chrysanthemum | 0–10 |
| Maidenhair fern | 15–18 |
| Kalanchoe | 10 |

*Suggested holding temperatures for cut foliage for at least short-term (<7 days) storage under high humidity (e.g. at least 90–95 per cent) conditions.

†RH in the range 65–90 per cent has been recommended for pot plants.

## TABLE 13.5  RECOMMENDED STORAGE TEMPERATURE AND STORAGE LIFE FOR 10 SELECTED BULBS*

| PLANT | (°C) | STORAGE PERIOD (MONTHS) |
|---|---|---|
| Alstroemeria | 4.5–10 | – |
| Caladium | 10–15.5 | – |
| Freesia | 22–30 | 3–4 |
| Gladiolus | 3.5–10 | 5–8 |
| Iris | 16–20 | 4–12 |
| Lily, Easter | -0.5–0.5 | 10 |
| Narcissus | 7–20 | 2–4 |
| Tulip, forcing | 4.5–10 | 2–4 |
| Tulip, outdoors | -0.5–0 | 5–6 |
| Zantedeschia | 2–4.5 | – |

*RH in the range 70–90 per cent is suitable for most bulbs. Air movement (ventilation) around dry stored bulbs is desirable.

However, if storage is necessary, a temperature of ≥13°C is generally recommended. Since the range of species of pot plants is substantial, some require storage at relatively low temperatures of about 5°C and some at high temperatures of about 20°C. Examples of recommended storage conditions for a range of foliage and potted plants is given in Table 13.4 and of bulbs in Table 13.5.

Many ornamental crops are very sensitive to ethylene (see Chapter 6), showing accelerated senescence or abscission. Ethylene-sensitive crops can be treated prior to storage with anti-ethylene compounds such as silver thiosulphate or stored in systems offering ethylene venting and/or scrubbing. Some examples of ethylene-sensitive cut flowers are given in Table 13.3.

## ATMOSPHERE CONTROL

The postharvest life of horticultural produce held at optimum temperature and humidity conditions can be further extended by control of the concentrations of carbon dioxide, oxygen and ethylene in the surrounding atmosphere. The optimum levels of carbon dioxide and oxygen are documented for many produce (see Chapter 6), but particular storage situations and marketing logistics will determine whether this is commercially feasible to implement. Recent studies have indicated that some produce may be sensitive to much lower concentrations of ethylene in the storage environment than previously believed, and thus more stringent removal of ethylene may be required. A re-evaluation of the storage effects of low concentrations of ethylene is required.

# FURTHER READING
·······

American Society of Heating, Refrigeration and Air-conditioning Engineers (1986) *ASHRAE handbook of refrigeration systems and applications*, ASHRAE, Atlanta, GA.

Anon. (1989) *Guide to food transport: Fruit and vegetables*, Mercantila Publishers, Copenhagen.

Hardenburg, R.E., A.E. Watada and C.Y. Wang (1990) *The commercial storage of fruits, vegetables, and florist and nursery stocks* (rev. ed.), US Department of Agriculture, Washington, DC. Handbook no. 66.

Jones, R. and H. Moody (1993) *Caring for flowers* (rev. ed.), Agmedia, Department of Agriculture Victoria, East Melbourne.

Ryall, A.L. and W.J. Lipton (1982) *Handling, transportation, and storage of fruits and vegetables* (rev. ed.), vol. 1, Vegetables and melons, AVI, Westport, CT.

Ryall, A.L. and W.T. Pentzer (1982) *Handling, transportation, and storage of fruits and vegetables* (rev. ed.), vol. 2, *Fruits and tree nuts*, AVI, Westport, CT.

Story, A. and D.H. Simons (eds) (1997) *Fresh produce manual*, Australian United Fresh Fruit and Vegetable Association Ltd, Sydney.

Thompson, A.K. (1996) Postharvest technology of fruits and vegetables, Blackwell Science, Oxford.

# Appendix I
# LIST OF
# ABBREVIATIONS

| | |
|---|---|
| 2-AB | secondary butylamine |
| ABA | abscisic acid |
| ACC | 1-aminocyclopropane |
| AFHB | Asean Food Handling Bureau |
| AHC | Australian Horticultural Corporation |
| ADP | adenosine diphosphate |
| AOA | aminooxyacetic acid |
| AQIS | Australian Quarantine Inspection Service |
| ATP | adenosine triphosphate |
| AVG | L-2-amino-4-(2-aminoethoxy)-trans-3-butenoicacid, aminovinyl glycene |
| BA | benzyladenine |
| $Ca^{2+}$ | calcium ion |
| CA | controlled atmosphere |
| $C_2H_4$ | ethylene |
| CIPC | 3-chloroisopropyl-N-phenylcarbamate |
| CO | carbon monoxide |
| $CO_2$ | carbon dioxide |
| 2,4-D | 2,4-dichlorophenoxyacetic acid |
| DNA | deoxyribonucleic acid |
| EDB | ethylene dibromide |
| EFE | ethylene forming enzyme |
| EMP | Embden-Meyerhof-Parnas (pathway) |
| ERH | equilibrium relative humidity |
| FAD | flavin adenine dinucleotide |
| $FADH_2$ | reduced form of FAD |
| FAO | Food and Agriculture Organisation (a United Nations Agency) |
| FDA | Food and Drug Administration (of the USA) |
| FDC | forced draft cooler |
| HACCP | hazard analysis and critical control points |
| HOPP | ortho-phenylphenol |
| IAEA | International Atomic Energy Authority (a United Nations Agency) |
| IDC | induced draft cooler |
| IICA | Inter-American Institute for Cooperation in Agriculture |
| ISO | International Standards Organisation |
| GA | gibberellins |

| Gy | Gray (unit of irradiation) |
| $KMnO_4$ | potassium permanganate |
| LPG | liquefied petroleum gas |
| MA | modified atmosphere |
| MAP | modified atmosphere packaging |
| MB | methyl bromide |
| MH | maleic hydrazide |
| mRNA | messenger ribonucleic acid |
| 1-MCP | 1-methylcyclopropene |
| MRL | maximum residue limit |
| NAD | nicotinamide adenine dinucleotide |
| NADH | reduced NAD |
| NIR | near infrared |
| $O_2$ | oxygen |
| OPP | ortho-phenylphenate (free anion) |
| PG | polygalacturonase |
| pH | log value of hydrogen ion concentration |
| Pi | inorganic phosphate |
| PIP | Postharvest Institute for Perishables |
| PME | pectin methylesterase |
| ppb | parts per billion |
| ppm | parts per million |
| QA | quality assurance |
| QC | quality control |
| QMS | quality management system |
| $Q_{10}$ | temperature quotient (10°C) |
| RH | relative humidity |
| RNA | ribonucleic acid |
| RQ | respiratory quotient |
| SA | surface area |
| SAM | S-adenosyl-methionine |
| $SO_2$ | sulphur dioxide |
| SOPP | sodium ortho-phenylphenate |
| SSC | soluble solids concentration |
| STS | silver thiosulphate |
| TA | titratable acidity |
| TBZ | thiabendazole |
| TCA | tricarboxylic acid (pathway) |
| TQM | total quality management |
| UDP-glucose | uridine diphosphoglucose |
| UV | ultraviolet |
| Vitamin C | ascorbic acid |
| V | volume |
| VP | vapour pressure |
| VPD | vapour pressure deficit |
| WHO | World Health Organisation (a United Nations Agency) |

# APPENDIX II
# GLOSSARY OF PLANT NAMES

**COMMON AND BOTANICAL NAMES OF SOME FRUITS AND VEGETABLES**

| COMMON NAME | BOTANICAL NAME |
|---|---|
| Apple | *Malus Xdomestica* Borkh. |
| Apricot | *Prunus armeniaca* L. |
| Asian pear | *Pyrus pyrifolia* Nakai and *P. bretschneideri* Rehder. |
| Asparagus | *Asparagus officinalis* L. |
| Avocado | *Persea americana* Mill. |
| Banana | *Musa* L sp., Cavendish varieties, *M. acuminata* Colla |
| Beans, broad | *Vicia faba* L. |
| string | *Phaseolus vulgaris* L. |
| mung | *Phaseolus aureus* Roxb. |
| Beetroot | *Beta vulgaris* L. |
| Blueberry | *Vaccinium* sp. |
| Broccoli | *Brassica oleracea* L. (Italica group) |
| Brussels sprout | *Brassica oleracea* L. (Gemmifera group) |
| Cabbage | *Brassica oleracea* L. (Capitata group) |
| Carambola | *Averrhoa carambola* L. |
| Carrot | *Daucus carota* L. |
| Cassava (manioc, tapioca) | *Mannihot esculenta* Crantz |
| Cauliflower | *Brassica oleracea* L. (Botrytis group) |
| Celery | *Apium graveolens* L. |
| Cherimoya | *Annona cherimola* Mill. |
| Cherry, sweet | *Prunus avium* L. |
| sour | *Prunus cerasus* L. |
| Chili | *Capsicum annuum* L. |
| Choko | *Sechium edule* (Jacq.) Sw. |
| Corn (maize), sweet | *Zea mays* L. |
| Cucumber | *Cucumis sativus* L. |
| Eggplant (Aubergine) | *Solanum melongena* L. |
| Feijoa | *Feijoa sellowiana* Berg. |
| Fig | *Ficus carica* L. |
| Garlic | *Allium sativum* L |
| Ginger | *Zingiber officinale* Rascoe |
| Globe artichoke | *Cynara scolymus* L. |
| Grape | *Vitis vinifera* L. |
| Grapefruit | *Citrus paradisi* Macfad. |
| Guava | *Psidium guajava* L. |
| Jackfruit | *Artocarpus heterophyllus* (Lam.) L. |
| Jerusalem artichoke | *Helianthus tuberosus* L. |
| Kiwi fruit | *Actinidia deliciosa* (A. Chev.) C.F. Liang et A.R. Ferguson |
| Leek | *Allium ampeloprasum* L, |

| | |
|---|---|
| Lemon | *Citrus limon* (L.) Burm. f. |
| Lettuce | *Lactuca sativa* L. |
| Lime | *Citrus aurantifolia* (Christm.) Swingle |
| Litchi | *Litchi chinensis* Sonn. |
| Loquat | *Eriobotrya japonica* Lindl. |
| Mandarin | *Citrus reticulata* Blanco |
| Mango | *Mangifera indica* L. |
| Mangosteen | *Garcinia mangostana* L. |
| Muskmelon (cantaloupe, honey dew) | *Cucumis melo* L. |
| Nectarine | *Prunus persica* (L.) Batsch. |
| Okra | *Hibiscus esculentus* L. |
| Onion | *Allium cepa* L. |
| Papaya (pawpaw) | *Carica papaya* L. |
| Parsley | *Petroselinum crispum* (Mill.) Nym. |
| Parsnip | *Pastinaca satiua* L. |
| Passionfruit | *Passiflora edulis* Sims |
| Pea | *Pisum sativum* L. |
| Peach | *Prunus persica* (L.) Batsch. |
| Pear | *Pyrus communis* L. |
| Pepino | *Solanum muricatum* Ait. |
| Peppers, green and red | *Capsicum annuum* L. |
| Persimmon | *Diospyros kaki* L.f. |
| Pineapple | *Ananas comosus* (L.) Merr. |
| Plum | *Prunus domestica* L. |
| Pomegranate | *Punica granatum* L. |
| Potato | *Solanum tuberosum* L. |
| Pumpkin | *Cucurhita pepo* L. |
| Radish | *Raphanus sativa* L. |
| Rambutan | *Nephelium lappaceum* L. var. *esculentum* Nees |
| Rhubarb | *Rheum* sp. |
| Satsuma mandarin | *Citrus unshu* Mari |
| Soya bean | *Glycine max* (L.) Merr. |
| Spinach, European | *Spinacia oleracea* L. |
| Squash | *Cucurbita maxima* Duch. |
| Strawberry | *Fragaria Xananassa* Duch. |
| Sweet orange | *Citrus sinensis* (L.) Osbeck |
| Swede turnip | *Brassica napus* L. (Napobrassica group) |
| Sweet potato | *Ipomea batatas* (L.) Lam. |
| Tamarillo (tree tomato) | *Cyphomandra betacea* (Cav.) Sendt. |
| Taro | *Colocasia esculenta* (L.) Schott |
| Tomato | *Lycopersicon esculentum* Mill. |
| Turnip | *Brassica campestris* L. (Rapifera group) |
| Watermelon | *Citrullus lanatus* (Thunb.) Mansf. |
| Yam | *Dioscorea batatas* Deene |
| Zucchini | *Cucurbita pepo* L. 'Zucchini' |

## COMMON AND BOTANICAL NAMES OF SOME CUT FLOWERS AND FOLIAGE

| | |
|---|---|
| Alstromeria, Lily of the Incas, Peruvian Lily | *Alstroemeria* spp. |
| Anthurium, Tailflower | *Anthurium andreanum* |
| Belladonna lily | *Amaryllis belladonna* |
| Bird-of-paradise, Strelitzia | *Strelitzia reginae* |

| | |
|---|---|
| Carnation | *Dianthus caryophyllus* |
| Cattleya orchid | *Cattleya* spp. |
| Christmas mistletoe | *Phoradendron tomentosum* |
| Chrysanthemum, Florist's, Mum | *Dendranthema* spp. |
| Cymbidium orchid | *Cymbidium* spp. |
| Delphinium, larkspur | *Delphinium* hybrids |
| Dendrobium orchid | *Dendrobium* spp. |
| Eucalyptus, Gum tree | *Eucalyptus* spp. |
| Freesia | *Freesia hybrida* |
| Gerbera, Transval daisy | *Gerbera jamesonii* |
| Ginger | *Alpinia* spp. |
| Grevillea | *Grevillea* spp. |
| Gypsophila, baby's breath, Gyp | *Gypsophila paniculata* |
| Iris, Bulbous iris | *Iris* hybrids, Dutch |
| Kangaroo paw | *Anigozanthos* spp. |
| Liatris, Gayfeather | *Liatris spicata* |
| Lily | *Lilium* spp. |
| Limonium, statice | *Limonium* spp. |
| Lisianthus, Texas rose | *Eustoma* grandiflorum |
| Narcissus, daffodil | *Narcissus* hybrids |
| Paphiopedilum orchid | *Paphiopedilum* spp. |
| Paper daisy | *Helipterum* spp. |
| Protea | *Protea* spp. |
| Rose | *Rosa* hybrids |
| Snapdragon | *Anthirrhinum majus* |
| Tulip | *Tulipa* hybrids |
| Vanda orchid | *Vanda* spp. |

## COMMON AND BOTANICAL NAMES OF POT PLANTS

| | |
|---|---|
| African violet, Saintpaulia | *Saintpaulia ionantha* |
| Azalea, Indian azalea | *Rododendron* spp. |
| Begonia | *Begonia* hybrids |
| Boston fern | *Nephrolepis exaltata* |
| Chyrsanthemum | *Dendranthema* spp. |
| Dracaena | *Dracaena* spp. |
| Ficus, Fig, Rubber tree | *Ficus* spp. |
| Frangipani, Plumeria | *Plumeria* spp. |
| Hibiscus | *Hibiscus* spp. |
| Ivy | *Hedera* spp. |
| Kalanchoe, Palm beach bells | *Kalanchoe* spp. |
| Melaeuca | *Melaleuca* spp. |
| Poinsettia | *Euphorbia pulcherrima* |
| Schefflera | *Schefflera polybotrya* |
| Spathiphyllum, Spathe flower | *Spathiphyllum* spp. |
| Yucca | *Yucca* spp. |

# APPENDIX III
# TEMPERATURE AND HUMIDITY MEASUREMENT

## TEMPERATURE
·······

The Celsius, or Centigrade, scale (°C) and the Fahrenheit scale (°F) are the main temperature scales used by horticulturalists. The Celsius scale is based on phase changes of water, with the freezing point of water being 0°C and the boiling point of water at atmospheric pressure being 100°C. Celsius is the standard international (SI) temperature scale. On the Fahrenheit scale, which has been widely used, the freezing and boiling points of water are at 32°F and 212°F, respectively.

### *Temperature-measuring devices*
A range of devices can be used to measure temperature. All of these devices need to be used with care if correct temperatures are to be measured, and all need calibration.

### *Liquid-in-glass thermometers*
These thermometers are the most commonly used temperature-measuring devices, and are based on the principle that a liquid expands when it is heated and contracts when cooled. The change in volume is read against a fixed scale. Liquid-in-glass thermometers are cheap, simple, easy-to-read instruments with an acceptable rate of response to changes in temperature, and their calibration changes little with time. However, they are fragile and must be handled with care; the sensing bulb or liquid reservoir is necessarily thin-walled and the *stem will not bend* — a common reason for breakage! Commonly used liquids are mercury and ethyl alcohol (coloured, usually red, for easy visibility); the freezing and boiling points are -38.9°C and 356.6°C for mercury and -115°C and 78.3°C for alcohol. Mercury is a poisonous substance and spilt mercury must be treated accordingly.

Liquid-in-glass thermometers will give incorrect readings if not properly constructed or used. Unless supplied with a standard calibration certificate, every thermometer should be checked at 0°C in an ice/water mixture and against a calibrated thermometer near the temperature at which it is to be used. Liquid-in-glass thermometers are designed to be either fully immersed or partly immersed in the material whose temperature is being measured, and are marked accordingly. When using thermometers, care should be taken that either extraneous heat from a hot body, such as your own, or a cold surface does not affect the reading. Sufficient time must also be allowed for the thermometer to come to equilibrium as the glass has some heat capacity and conductance.

## Fluid-filled dial thermometers

These thermometers enable measurements to be made at a distance of a few metres, and therefore they are often used as externally indicating, dial thermometers in cool stores. They consist of a sensing bulb connected by a capillary to a spiral tube, called a Bourdon tube, and the whole system is filled with a liquid, a gas, or a saturated vapour. The pressure changes caused by varying temperature cause movement of the Bourdon tube, which is linked to a pointer that indicates the temperature on a graduated circular dial. These thermometers are constructed with small bulbs made from a material such as copper, which has a low specific heat and is highly conductive. Response to changes in temperature is slower than that of liquid-in-glass thermometers. These thermometers should be calibrated annually.

## Bimetallic thermometers

These thermometers comprise a laminate of two metals, one with a high and one with a low coefficient of expansion. The laminate changes shape with temperature variation and activates a pointer or dial; the response to changes in temperature is slow. The indicated reading is subject to small errors because of friction in the mechanical components. These errors can be partially overcome by gently tapping the dial. Bimetallic thermometers are robust and are convenient for measuring internal flesh temperatures of produce by inserting the pointed measuring tip. To minimise errors caused by heat conducted along the metal stem, the thermometer should be immersed to the depth indicated, or to at least 20 times the diameter of the stem.

## Thermographs

A thermograph combines a bimetal thermometer with a moving paper chart to give a temperature record. The chart is driven by clockwork or electric battery.

## Thermocouples

If two strips of different metals are joined together at each end and the junctions are kept at different temperatures, an electromotive force (emf) develops in the two-part conductor. This electromotive force depends on the metals and the temperature difference. This principle is used in most thermoelectric thermometers. The simplest form comprises the thermocouple of two wires soldered together at their ends with a meter in the circuit to measure the generated electromotive force. One junction is kept at a known reference temperature, commonly at 0°C in melting ice, and the other is the sensing element placed where the temperature is to be measured. Alternatively, one end can be set at an 'electronic zero', obviating the need for temperature control at the reference end (junction). A commonly used pair of metals is copper and constantan (an alloy of 60 per cent copper and 40 per cent nickel) and gives an electromotive force of about 39 μV/°C.

Thermocouples have the following advantages:

◆ the length of the wires is not significant so that they can be used in remote positions;

◆ the measuring probe can be small; and

◆ high precision and rapid responses can be obtained.

Readings can be taken manually or recorded automatically. Thermocouples are most useful for studying temperature changes in space and time, such as in a filled cool store or a loaded transport vehicle.

## Digital thermometers

A digital thermometer uses an electronic circuit to detect the output of a thermocouple, resistance element or thermistor, converts this to a temperature and shows the reading on a digital display. Although the temperature displayed may be up to 0.1°C, digital thermometers are not necessarily as accurate as liquid-in-glass thermometers. Both hand-held battery-operated and panel-mounted mains-powered digital thermometers are in use. Panel-mounted digital thermometers, coupled with a resistance sensor (temperature-sensitive wire, usually platinum, or a thermistor) are commonly used for permanent installations, such as in refrigerated ships. Digital thermometers using a resistance sensor do not require a cold junction. The accuracy of hand-held thermometers used in the field should be checked frequently in melting ice at 0°C.

## Data-loggers

As an extension of the digital thermometer, the data-logger measures temperatures at preset intervals of time, and stores the values in solid-state memory for later recall and analysis. Battery and mains-powered models are available. Relatively inexpensive single point data-loggers are available for recording temperatures in packaged produce during commercial shipment, while more sophisticated multi-point data-loggers are available for research purposes.

## Calibration of thermometers

Melting ice, made from potable water, has a temperature within 0.05°C of 0°C. Preferably, ice made from distilled and/or deionised water should be used. For checking at 0°C an ice/water mixture should be prepared in an insulated container such as a 'Thermos' flask rather than in a beaker or glass jar. Another method is to use wet crushed ice and to tamp the ice down gently in a shallow container to obtain a 'solid' mass. The surface of the ice should look wet. The stem of the thermometer to be tested should be gently forced into the ice. Alternatively, a thick ice and water slurry can be used, but this should be well stirred. For checking at other temperatures, a calibrated mercury-in-glass thermometer with 0.1°C graduations should be obtained. A thermometer with a calibration certificate may be purchased, or a thermometer may be submitted to an approved testing laboratory for calibration.

## Where to measure temperature

The placement of temperature-measuring devices in cool stores and similar structures is important, particularly when the rates of cooling (or warming) of produce are being followed or when refrigeration or heating is being controlled by thermostats. The nature of the refrigeration system means that there must always be a gradient in temperature between the produce and the cooling coils. In addition, the gradient is accentuated by excessive heat leakage through the structure or through faulty stacking of packages. Generally the aim is to maintain the bulk of the stow at the recommended storage temperature without freezing or over-cooling the coldest part. The device that is used as a thermostatic control for the forced draft cooler is best placed in the air coming off the cooler, with the thermostat adjusted to give the minimum acceptable temperature.

Temperature of produce in a cool store, ship's hold or container should be measured in several different positions because there will inevitably be spatial variations that may be enough to adversely affect the cooling of some of the load. Air temperatures just inside the door never give accurate readings of produce temperature.

# HUMIDITY
·······

The amount of water vapour in air (i.e. the psychrometric state of the atmosphere) can be specified either by the water content or by its vapour pressure, and either in absolute or relative terms. Relative humidity (RH) is the ratio of water vapour pressure in air to saturation vapour pressure at the same temperature, expressed as a percentage.

RH = $(P/P_0)T$ x 100%

where P = water vapour pressure of air at temperature T

$P_0$ = saturation vapour pressure at the same temperature T

It is important to remember that relative humidity can only be compared at the same temperature and barometric pressure.

Absolute (or specific) humidity is the measure of the weight of water vapour contained in a known weight of dry air. Typical psychrometric charts provide a scale showing absolute humidity in grams or kilograms of water vapour per kilogram of dry air. Absolute humidity is proportional to vapour pressure (Fig. 5.6). The absolute humidity of saturated air at 10°C, 20°C and 30°C is approximately 8 g/kg, 15 g/kg and 27 g/kg, respectively.

Saturation vapour pressure is the vapour pressure of water in equilibrium with a free water surface. Alternatively it can be defined as the pressure exerted by the maximum amount of water that can be contained in the air at a given temperature. The saturation vapour pressure increases rapidly as the air temperature rises.

Dew point is the temperature at which saturation occurs when air is cooled without change in water content. This is also a practical parameter as it specifies temperature and 100 per cent relative humidity simultaneously, and hence saturation vapour pressure or saturation water content as well. Changes in air temperature above the dew point do not affect the water content, but cooling below the dew point removes moisture from the air by condensation on cooler surfaces. Moisture will condense on any surface cooler than the dew point of the air in contact with it.

## *Types of hygrometers*

An instrument use to measure the amount of water in air is termed a hygrometer or psychrometer. Just as there are several ways of defining the amount of water in air, many methods have been devised for its measurement. No single hygrometer (or psychrometer), however, is suitable for all purposes over the full range of humidities and temperatures.

## Wet and dry bulb hygrometers

This is the simplest and most widely used instrument for measuring humidity. It consists of two thermometers, one of which — the dry bulb — measures the air temperature. The wet bulb thermometer has a wet wick around the bulb. Evaporation of water from the wick into the atmosphere requires energy, which comes from the remaining water and results in it being cooled. The drier the air, the greater is the rate of evaporation and hence the greater the depression of temperature. The temperature depression can be translated to per cent of relative humidity, water vapour pressure or dew point from tables prepared for this purpose. The values in the tables vary with atmospheric pressure, but the effect can be ignored for most practical purposes in the range 82–101 kPa (620–760 mm mercury), which is equivalent to altitudes up to 1500 metres.

Certain precautions are necessary for accurate readings. The wick must be clean and free from dust and other contamination. Use of distilled water only is recommended. The bulbs must be ventilated with air moving at least 3 m/sec to ensure adequate evaporation and cooling of the wet bulb. The formula used to produce the tables assumes adiabatic evaporative cooling only, so the instrument should be protected from radiation sources such as the Sun, light bulbs or any surface much colder or warmer than the surrounding air. With careful operation and accurate thermometers reading to 0.1°C, an accuracy of ±1 per cent relative humidity is possible. The simplest is the sling (or whirling) psychrometer, while in the Assmann type, aspiration is by means of a small motorised fan. The latter is the standard instrument as it also has a built-in metal radiation shield and has finely calibrated and accurate thermometers. For general use, a sling psychrometer is adequate and most practical. Wet bulb and dry bulb instruments are now available with thermistors instead of mercury-in-glass thermometers; their advantages are small size, the possibility of automatic and distant operation, and remote control.

## Hair hygrometers

The sensing element of these hygrometers comprises several strands of hair, or a length of some other material that is capable of water sorption and desorption with a consequent change in length. The sensing element is mechanically linked to a pointer on a scale. As the capillary sorption of water is slow, the response of these hygrometers is slow (10–30 minutes) and, as it also depends to some extent on the amount of water in the system, there are marked hysteresis effects. Therefore, the instruments should not be exposed to conditions of widely fluctuating temperatures or humidities. They must be calibrated for each temperature range. Over the range 30–80 per cent relative humidity they have an accuracy of 2–5 per cent relative humidity. They are useful for monitoring slow humidity variations at almost steady temperatures, as in cool stores.

### Electric hygrometers

These instruments measure the psychrometric state of the atmosphere by recording variations in the resistance, capacitance or some other electrical parameter of a sensor with changing water sorption or desorption. Metal or carbon electrodes are attached to an insulating base impregnated with, or covered with a thin layer of a dilute solution of an electrolyte, which equilibrates with the surrounding air by sorbing or desorbing water. The conduction types are temperature-dependent and suffer from hysteresis and ageing effects, but these instruments are small and have a fast response. The signals can be amplified and these types can be used readily for remote control. Capacitor sensors are more stable and reliable. They are unstable at high humidities, and must not be allowed to get wet.

### Thin-film polymer hygrometers

Like some electric hygrometers, these measure changes in capacitance of a material that is sensitive to humidity. The sensor consists of a thin solid solution, which makes these instruments more robust and stable. Some can be washed with distilled water without losing calibration.

### Dew point hygrometers

To measure the psychrometric state of the atmosphere with this type of hygrometer, air is cooled without change in water content until saturation is reached. The temperature (dew point) at which condensation is first seen on a cooled mirror surface is recorded, and from this temperature the vapour pressure or relative humidity of air can be derived.

# FURTHER READING
·······

American Society of Heating, Refrigerating and Air-Conditioning Engineers (1972) *ASHRAE handbook of fundamentals*, ASHRAE, New York, chapters 5 and 12.

Gaffney, J.J. (1985) Humidity: Basic principles and measurement, *HortScience* 13: 551–55.

Grierson, W. and W.F. Wardowski (1975) Humidity in horticulture, *HortScience* 10: 356–60.

Hall, J.A. (1953) *Fundamentals of thermometry*, Institute of Physics, London. Monographs for students.

Henry, Z.A. (ed.) (1975) *Instrumentation and measurement for environmental sciences*, American Society of Agricultural Engineers, St. Joseph, MI.

Middlehurst, J. (1964) Temperature measurement, *CSIRO Food Preserv. Quart.* 24: 5–10.

Schurer, K. (1985) Comparisons of sensors for measurement of air humidity, in D. Simatos and J.L. Multon (eds) *Properties of water in foods*, Martinus Nijhoff, Dordrecht, pp. 647–60.

Sharp, A.K. (1986) Humidity: Measurement and control during the storage and transport of fruits and vegetables, *CSIRO Food Res. Quart.* 46: 79–85.

Szulmayer, W. (1969) Humidity and moisture measurement, *CSIRO Food Preserv. Quart.* 29: 27–35.

# APPENDIX IV
# GAS ANALYSIS

Work with fresh produce frequently requires measurement of the concentrations of carbon dioxide, oxygen and ethylene. These gases can be measured by both chemical and physical methods. However, the latter methods are preferred because of their speed, accuracy and sensitivity. The method chosen depends on the available gas sample size, the level of accuracy required, and whether continuous monitoring of gas concentration is necessary. Some techniques commonly used for gas analysis in postharvest work are briefly described.

## Orsat gas analyser

The Orsat gas analyser is used for determining the concentrations of carbon dioxide and oxygen in large gas samples, such as those from controlled atmosphere storage rooms. The analyser consists of a calibrated burette connected by a glass manifold to two absorption tubes; the first tube is filled with potassium hydroxide to absorb carbon dioxide, the second with alkaline pyrogallol or ammoniacal cuprous chloride to absorb the oxygen. The concentration of carbon dioxide and oxygen is determined by the reduction in volume of the air sample as measured in the burette. An accuracy of 0.1 per cent is possible with the analyser. Smaller, robust propriety variants of the Orsat have been developed.

## Thermal conductivity gas chromatography

Thermal conductivity gas chromatography is particularly suited to the analysis of small samples (0.2–5 mL). Particular applications include analyses of carbon dioxide and oxygen concentrations in small containers such as plastic bags, where the taking of samples of the size required for an Orsat gas analyser would drastically change the composition of the atmosphere in the bags. Gas samples are generally fractionated into the component gases oxygen, nitrogen and carbon dioxide on a silica gel column followed by a molecular sieve column. The concentration of each component emerging from the columns is determined by a detector responding to variations in the thermal conductivity of the separated gases passing over the detector. The carrier gas used in this application is helium.

## Infrared analyser

This instrument measures carbon dioxide in flowing gas streams and is especially suitable for incorporation into automatic systems. Carbon dioxide absorbs infrared radiation at a specific wavelength, a property that is used to produce an electrical signal related to the carbon dioxide concentration in the test gas stream. Commercial instruments are not flow-sensitive. This instrument can also be used in a similar way to a gas chromatograph. In this application, nitrogen is passed at a constant rate through the detector and samples of up to 5 mL (pulse) are injected into the nitrogen stream. Peaks proportional to the concentration of carbon dioxide in the sample are recorded on a strip chart recorder or by electronic integration of the peak area.

## Paramagnetic analyser

This instrument measures oxygen concentrations in flowing air streams. The paramagnetic, or magnetic susceptibility, analyser is limited to the analysis of oxygen and the oxides of nitrogen since these are the only paramagnetic gases (i.e. gases attracted by a magnetic field). Paramagnetic analysers can determine differences in oxygen concentrations of 0.01 per cent. A paramagnetic analyser can be connected in series with an infrared gas analyser to measure both oxygen and carbon dioxide concentrations simultaneously in either a flowing system or in injected samples as described above for carbon dioxide.

## Ethylene determination by gas chromatography

Gas chromatography using a flame ionisation detector is a common, highly sensitive system for the measurement of ethylene and other hydrocarbons up to C5. The ethylene in the gas sample is separated from the other gases on an alumina column; the separated ethylene emerging from the column is mixed with hydrogen, burnt in air, and the ions given off in the flame provide an electrical signal proportional to the amount of ethylene present in the gas sample. The carrier gas is high purity nitrogen. Commercial instruments vary widely in sophistication, but a single column, single detector instrument is adequate for routine ethylene analysis. The more sophisticated research instruments can measure 0.001 µL/L ethylene in a 5 mL sample of air.

Highly sensitive gas chromatographs fitted with a photoionisation detector have been developed. Carrier gas (usually nitrogen) only is required, and the separated gases emerging from the column are fed directly to the detector. Both laboratory and portable versions of this instrument are available commercially.

Two infrared laser-driven systems have been developed for *in situ* measurement of ethylene production by plants. A photothermal detector has a lower limit of detection of 0.5 nL/L while a photoacoustic detector can detect concentrations as low as 6 pL/L.

# FURTHER READING
·······

De Vries, H.S.M., F.J.M. Harren and J. Reuss (1995) *In situ*, real time monitoring of wound induced ethylene in cherry tomatoes by two infrared laser-driven systems, *Postharvest Biol. Technol.* 6: 275–85.

Dilley, D.R., D.H. Dewey and R.R. Dedolph (1969) Automated system for determining respiratory gas exchange of plant materials, *J. Am. Soc. Hortic. Sci.* 94: 138–41.

McNair, H.M. and E.J. Bonelli (1969) *Basic gas chromatography*, Varian, Walnut Creek, CA.

Thompson, B. (1977) *Fundamentals of gas analysis by gas chromatography*, Varian, Walnut Creek, CA.

Watada, A.A. and D.R. Massie (1981) A compact automatic system for measuring $CO_2$ and $C_2H_4$ evolution by harvested horticultural crops, *HortScience* 16: 39–41.

Young, R.E. and J.B. Biale (1962) Carbon dioxide effects on fruit respiration: Measurement of oxygen uptake in continuous gas flow, *Plant Physiol.* 37: 409–15.

# INDEX

abscisic acid (ABA) 95, 191, 192
abscission 188
absolute humidity 250
acetaldehyde 97
acidifiers 95
acidity 175–76
active packaging 226
aerobic respiration 38, 47, 48
AHC (Australian Horticultural
    Corporation) 182
air-cooled stores 114–15
air movement 91
air pressure reduced 91
air-wash refrigeration systems 71
*Alternaria* 144, 152, 155, 156
*A. alternata* 166
aminoethoxyvinylglycine (AVG)
    110
aminooxyacetic acid (AOA) 110
anaerobic respiration 38, 48, 52
anthocyanins 55
anthracnose 153
avocadoes *colour plate,* 166
    mangoes 145, 155
    papayas 145
antioxidants 202–203
AOA *see* aminooxyacetic acid
appearance 160–63
apples
    aroma 29
    calcium application 204–205
    composition 30
    disorders 136–38, 142, 153
    effect of VPD on weight loss 89
    ethylene in 40
    half-cooling times 70
    injury 219
    radiation 202–203
    respiration 232
    storage life 232
    transpiration coefficients 86
    vitamins in 27
apricots
    injury 219
    radiation 202–203
    storage life 232
    vitamins in 27
aroma 29, 30, 56, 163
artichokes, vitamins in 28
ascorbic acid 1, 25–26
    loss 78
asparagus
    radiation 202–203
    storage life 232
    vacuum cooling 74
    vitamins in 28

atmosphere
    modified and controlled
        11–12, 238
    storage 97–112
    *see also* controlled atmosphere;
        modified atmosphere
atomic oxygen for removal of
    ethylene 110
Australian Horticultural
    Corporation
    *see* AHC
auxins 191
AVG *see* aminoethoxyvinylglycine
avocadoes 165
    chilling injury 131, 132
    disorders 142
    ethylene in 40
    radiation 202–203
    ripening conditions 209
    storage life 232
*Bacillus subtilis* 156
bacteria 144
bananas
    aroma 29
    chilling injury 131
    composition 30
    disorders 153
    ethylene in 40
    injury 219
    modified atmosphere 102
    physiochemical changes during
        ripening 37
    radiation 202–203
    respiration 233
    ripening 43–44, 205–209
    shelf life 109
    storage life 232
    use of ethylene in ripening
        106
    vitamins in 27, 28
batch degreening 210
beans
    disorders 142
    storage life 232
beetroot, vitamins in 27
benomyl 150, 152, 154
bent neck of roses 95
benzimidazoles 150, 152, 154,
    155
berries, storage 232, 236
bimetallic thermometers 247
bin storage 220
biocides 95
biological control 156–57
biphenyl 152
bitter pit *colour plate,* 138,
    141–42, 204

black currant, vitamins in 27
black heart in pineapple *colour
    plate,* 133
black rot 153
blanket stores 122
blossom-end rot of tomatoes
    141
blue mould *colour plate,* 149,
    153
blush colour 173
boiled condition of tomato and
    banana 64
boron deficiency 141–42
*Botrytis* 144, 145, 149
*B. cinerea* 82
    rots in lettuce 111
broccoli
    composition 30
    storage life 232
    vacuum cooling 74
    vitamins in 27, 28
brominated charcoal 109
brown core 138
brown heart 138, 139
browning conditions 136, 137,
    140
    apples and pears 137
    plums 133
    through mechanical damage
        216
brown rot 153
    peaches *colour plate,* 155, 156
bruising 26
Brussels sprouts
    disorders 142
    transpiration coefficients 86
    vitamins in 27, 28
bulbs, recommended storage
    temperature 239
bulbs, roots and tubers, vegetable
    57
cabbages
    aroma 30
    disorders 142
    storage life 232
    transpiration coefficients 86
    vacuum cooling 74
    vitamins in 27, 28
calcium 26, 28
    application 204–205
    deficiency disorders 141, 142
calendar date 171–72
cantaloupes
    injury 219
    ripening conditions 209
carbendazim 152
carbohydrates 22–24

carbon dioxide 97–106
  control in packaging 227
  increase in 100–102
  measurement 254–55
carbon monoxide 11
carnations
controlled atmosphere 99
  effect of temperature on 62
  genetically engineered 110–11
carotenoids 54
carrots
composition 30
  disorders 142, 153
  storage life 232
  transpiration coefficients 86
  vacuum cooling 74
  vitamins in 27, 28
cassava
  storage of 113, 114
  vitamins in 27
cauliflower, vitamins in 28
celery
  disorders 142
  storage life 232
  vacuum cooling 74
cellars 113–14
cells *see* plant cells
centre rot 156
chemical changes during
  maturation 54–58
chemical measurements 175–77
chemical residues 167–68
  hazard control 185
chemicals
  as fungicides 148–55
  in vase solutions 95
cherries
  disorders 142, 153
  storage life 232
chicory, disorders of 142
chilli, vitamins in 27
chilling injury *colour plate,* 64,
  130–35, 138–39
chilling-sensitive produce, effect
  of temperature on 60, 62
chlorine 149
chlorophyll degradation 54, 55,
  102, 210
chlorophyll fluorescence
  spectrophotometers 177–78,
  179
CIPC (3-chloroisopropyl-N-
  phenylcarbamate) 192–93
citric acid 25, 50
citrus
  chemical treatments 149, 152,
  154–55, 156
  colour 209–10
  disorders 139, 153

maturity 176, 209–10
  storage 237
  vitamins in 27
  *see also* names of specific fruit
    e.g. orange; grapefruit
clamp storage 113, 114
climacteric behaviour 176–77
climacteric fruit 38–39
  ethylene and 40, 41
cobalt 142
coconuts, storage life of 232
cold treatments 196–97
*Colletotrichum* 144, 145, 146
*C. gloeosporioides colour plate,* 166
colour 161–62, 173–74
  fruit 54–55
  measurement of 173–74
  ornamentals 58
Colour Add 210
commercial maturity 168–69
  determining 170–79
compression bruising 216
condensation 72, 84–85, 92, 233
  in packages 226
condition and defects 162–63
*Coniothyrium* 156
contact icing 73
controlled atmosphere storage
  52–53, 97, 98–106
  for cut flowers 99
controlled atmosphere stores
  design and construction 120–23
  management of 123–27
  safety 123
controlled degreening 209–10
controlled ripening 205–209
cooling 66–75
  in package 219–20
cooling rates 68–69
cool stores, design and
  construction 117–19
copper 142
core flush *colour plate,* 137, 138
corn *see* sweet corn
critical temperatures 134
crown rot in bananas *colour plate,*
  145, 153, 155
cucumbers
  aroma 30
  chilling injury 131
  storage life 232
curing 88, 146, 191, 236
cut flowers 17
controlled atmosphere conditions 99
  packing 223
  pulsing and desiccation 211–12
  storage conditions 234, 235, 237
  structure 19
  *see also* ornamentals

cuticle 87
cytokinins 191, 192
*Debaryomyces* 157
deciduous tree fruit storage 236
deep scald 138
degree days 172
degreening, controlled 209–10
desiccation 211–12
deterioration of produce 6–7
development phase 34
dew point 84, 85, 250
dew point hygrometers 252
  2,4-dichlorophenoxyacetic acid
  192
dicloran 150, 155
dietary fibre 2, 24
digital data-loggers 249
digital thermometers 248
diphenylamine 202–203
*Diplodia* 144, 152
direct expansion system 116
diseases of affluence in humans
  24
diseases of postharvest produce
  153
disinfestation 193, 196, 198–99
disorders, physiological 130–43
dormancy 192
*Dothiorella dominicana colour plate*
  192
drying fruit 211
dry rot 153
economic importance of
  postharvest loss 8
EDB (ethylene dibromide) 193
edible flowers, buds, stems and
  leaves 57
eggplants, chilling injury of 131
electrical characteristics 175
electric hygrometers 252
electron beam radiation 197, 200
electron transport system 49, 50
EMP (Embden-Meyerhof-Parnas)
  pathway 47
*Enterobacter cloacae* 156
environmentally sound packaging
  227–28
enzyme dissociation 135
enzymic browning 142
ERH (equilibrium relative
  humidity) 81, 82
*Erwinia* 144
escarole, disorders of 142
ethanol 97, 139, 157
ethephon 210
ethoxyquin 202–203
ethylene 97, 102, 106–107,
  191–92, 238
  biosynthesis 42
  effect on cut flowers 58

effect on fruit 40–45
  measurement 255–56
  methods of reducing 108–11
  mode of action 42–44
  use in ripening 205–206, 209
ethylene scrubbers 226–27
evaporation 86
evaporative cooling 74–75
external generators 120–21
fermentation 38, 52
fibreboard packaging 217, 222
figs, storage life 232
flavour 164
flesh breakdown in apples 137
flesh firmness 174
flowers *see* cut flowers
fluid-filled dial thermometers
  247
foliage storage recommendations
  237, 239
folic acid 1, 26, 27
forced air cooling 70–72, 207,
  208
freesia, controlled atmosphere
  99
freezing injury 63–64, 131
fruit
  chemical composition 21–32
  defined 15
  physiology and biochemistry
    33–59
  ripening 34, 35–46, 54–56
  structure 15, 16
  world production 3–5
fruit flies 193, 194–95
  disinfestation 196–97
fruit pressure testers 174–176
fruit vegetables *see* vegetable
    fruits
fumigation 193, 196
fungal decay in cucumber *colour
    plate*
fungi 144
fungicide resistant microbes
  147, 154–55
fungicides 92, 147, 150–52,
  154–55

gamma radiation 197, 200
garlic storage 115
gas analysis 254–56
gas chromatography 254,
  255–56
gas separators 121
gas storage 97
genetic engineering
  of carnations 110–11
  of plant metabolism 52–53
  of ripening 45–46
*Geotrichum* 152, 155

*G. candidum  colour plate,* 156
gibberellins (GA) 191, 192
global trade 180
glycolysis 47–49
grading 212
grapefruit
  aroma 30
  storage life 232
  transpiration coefficients 86
grapes
  disorders 139, 153
  injury 219
  radiation 202–203
  respiration 233
  storage life 232
green colour 54, 173
green-life 177
green mould *colour plate,* 149,
    153, 156, 157
grey mould 153, 155
  of strawberries 147, 148
ground colour 173
growth 33–34
guava, vitamins in 27
guazatine 154
HACCP (hazard analysis and
    critical control points)
    181–85
hair hygrometers 251–52
half-cooling time 68, 70, 71
handling and transport 166–67
Harrison, James 115
harvesting 166, 188–90
  timing 165, 209
hazard analysis and critical
    control points *see* HACCP
heat treatment 148, 155–56,
    197, 200
heat units 172
hexanal 157
high temperature injury 64
hollow fibre membrane systems
    121, 122
hot water dips 148
humectants 95
humidity 82, 233–34
  increasing 90–91
  measurement 250–52
  and water loss 77–96
hydration 211
hydrocooling 72–73
hygrometers 250–52
hypobaric storage 105–106
ice refrigeration 115
ice slurry 73
icing 73
imazalil 150, 154
impact bruising 216
infection processes 144–46

infrared analyser 255
infrared spectroscopy 178
in-ground storage 113–14
injury *see* mechanical injury;
    chilling injury; disorders;
    diseases
insecticides 196
insect pests 194–95
internal cork 141–42
international trade 5, 6, 7
ionising radiation *see* irradiation
iron 26, 142
irradiation 168, 197, 200–203
isocitric acid 25
jacketed systems 122–23
Jonathan spot 137, 138
kangaroo paws, packaging and
    transport 95
Kidd and West 98
kiwifruit
  ripening conditions 209
  vitamins in 27
leaf cross-section 87
leafy vegetables
  disorders 153
  storage conditions 234
  vacuum cooling 73–74
leeks
  storage life 232
  transpiration coefficients 86
lemons
  aroma 30
  chilling injury 131
  ethylene in 40
  radiation 202–203
  respiration 233
lenticel rot 153
lenticels 88
lettuces
  composition 31
  disorders 111, 142
  storage life 232
  transpiration coefficients 86
  transport 98
  vacuum cooling 74
  vitamins in 27, 28
light processing 210–11
lily, controlled atmosphere 99
Lima beans, vitamins in 28
limes 1, 163
  chilling injury 131
  ethylene in 40
lipid hypothesis of chilling 135,
    234
lipids 25
liquid-in-glass thermometers
    246–47
liquid nitrogen refrigeration
    104–105

*Listeria* 211
loading 124
low temperature breakdown 138
magnetic resonance systems 178
Maillard reaction 65
maleic hydrazide *see* MH
malic acid 25, 50
mandarins, storage life 232
mangoes
  anthracnose in 145, 155
  chilling injury 131
  composition 31
  disorders 142, 153
  ethylene in 40
  green life and weight loss 77
  ripening conditions 209
  storage life 232
  vapour pressure deficit in 80
  vitamins in 27
  VPD and 80
marketing 6–7, 167
  flexibility in 230
market preparation 188–213
marrows, storage life 232
maturation 34
  chemical changes during
    54–58
maturity, determining 168–79
maturometer 170
maximum residue limits (MRL)
    155, 204
MB (methyl bromide) 193
mealybugs 194
mechanical injury 88–89, 166
  prevention of 216–18
  susceptibility of produce 219
melons
  chilling injury 131
  ripening conditions 43, 44, 209
mesh bags 92, 218, 223
metabolic stress 165
metabolism
  effect of temperature on 61
  effects of controlled
    atmosphere on 100–102
metabolites for synthetic reactions
    52
methyl bromide *see* MB
1-methylcyclopropene (1-
    MCP) 110
MH (maleic hydrazide) 192, 193
microbes 144
  effect of modified atmopheres
    on 102–103
  effects of ERH on 81, 82
  effects of temperature on 65
  fungicide resistance 147, 154–55
  growth 233
  microbial damage 166

mimosa, controlled atmosphere 99
mineral deficiency disorders
    140–42
minerals and vitamins 25–28
minimal processing 13, 210–11
modified atmosphere packaging
    97
modified atmosphere storage
    52–53, 97, 98–106
  effect on chilling 134
molecular biology 13
molecular probes 178
*Monilinia* 144
*M. fructicola colour plate*
mouthfeel 163
mushrooms
  aroma 30
  radiation 202–203
  storage life 232
nectarines
  ethylene in 40
  injury 219
  radiation 202–203
nitrogen 104–105
nitrogenous compounds 56
non-chilling-sensitive produce
  effect of temperature on 60,
    62–63
non-climacteric fruit 39
  ethylene and 40, 41
nutritional value 1–3, 164–65
oleocellosis 188
onions
  sprouting in 192–93
  storage 115
  storage life 232
  transpiration coefficients 86
  vacuum cooling 74
  vitamins in 27, 28
OPP (*o*-phenylphenate anion)
    149, 152
OPPP (oxidative pentose
    phosphate pathway) 50, 51
oranges
  aroma 30
  ethylene in 40
  harvesting 188
  maturity 173, 176, 177
  radiation 202–203
  respiration 233
  storage life 232
  vitamins in 28
orderly marketing 126
organic acids 25, 56
ornamentals
  anti-ethylene treatments 110–11
  effect of ethylene on 107
  maturation 58
packing 223

physiological development 35
  quality testing 179
  reducing water loss 93–95
  storage 237–38, 239
  use of 2–3
  world production 5–6
  *see also* cut flowers
Orsat gas analyser 254
osmotic potential 79
over-storage 126
oxalic acid 25
oxidative pentose phosphate
    pathway *see* OPPP
oxidative phosphorylation 50
oxidisation of ethylene 108–10
oxygen 97–106
  atmospheric 52
  control in packaging 227
  decrease in 100–102
  measurement 254–55
  ozone 109
package dimensions 220–21
package strength 222
package testing 222–23
packaging 12, 92–93, 214–29
packing 223–24
*Paecilomyces colour plate,* 156
papayas
  anthracnose in 145
  chilling injury 131
  disorders 153
  radiation 202–203
  ripening 173
  ripening conditions 209
  vitamins in 27
paramagnetic analyser 255
parsley
  composition 31
  vitamins and minerals in 26,
    27, 161
parsnips
  disorders 142
  transpiration coefficients 86
partial pressure infiltration
    204–205
passionfruit
  ethylene in 40
  storage life 232
pathology 144–57
pattern packing 223
peaches
  disorders 139, 153
  ethylene in 40
  injury 219
  radiation 202–203
  respiration 233
  storage life 232
  transpiration coefficients 86
  vitamins in 27

pears
  disorders 139, 142, 153
  ethylene in 40
  injury 219
  quality in 159
  radiation 202–203
  respiration 233
  ripening conditions 209
  ripening temperature 62
  storage life 232
peas
  storage life 232
  vacuum cooling 74
  vitamins in 28
penetrometers 174
*Penicillium* 144, 147, 149, 152
*P. digitatum colour plate,* 144, 146, 149, 152
*P. expansum* 144
*P. frequentans* 156
*P. italicum colour plate,* 149
peppers, disorders 142
periderm 88
persimmons, ripening conditions 209
pests, effects of temperature on 65
pH levels 146
*Phlyctaena vagabonda* 145
*Phomopsis* 144
physical loss 8
physiological disorders 12, 130–43
physiological maturity 168
phytohormones 191–92
*Pichia* 157
pigment *see* colour
pineapples
  chilling injury 131
  disorders 133, 153
  ethylene in 40
  maturity 173
  storage life 232
  vitamins in 27
pit storage 113
place packing 223
plant cells 17–21
plant growth regulators 191–92
plant tissue 78–81
plastic film 92, 224–25, 227
  use in modified atmosphere storage 105
plastic tents 122–23
plums
  disorders 139
  ethylene in 40
  injury 219
  storage life 232
polyethylene 92, 105, 224–25
polystyrene packaging 218

polyurethane, foamed-in-place 123
postharvest infection 145–46
postharvest losses 8, 9–10
postharvest technology 6–8, 10–13
postharvest wastage control 147–57
potassium 26, 142
potassium permanganate 108–109, 226
potatoes 191
  aroma 30
  composition 31
  disorders 142, 153
  greening 102
  radiation 202–203
  sprouting in 192–93
  storage 65, 115, 232
  transpiration coefficients 86
  vacuum cooling 74
  vitamins in 27, 28
potted plants
  packing 223
  potting mix 93–94
  storage recommendations 237–38, 239
  *see also* ornamentals
precooling 67, 123
preharvest infection 145
preharvest wastage control 147
prepacking, consumer 224
pressure cooling 70–72, 219
pressure swing adsorption machine 121
prochloraz 154–55
protein 24
protopectin 55
provitamin A carotenoids 1
*Pseudomonas* 144
psychrometer 250
pulses, vitamins in 27
pulsing 211–12
pumpkins, storage life 232
QA (Quality Assurance) 181, 182
QC (Quality Control) 181, 182
quality
  evaluation and management 12, 159–87
  loss 8
Quality Management Systems 181–84
radishes, aroma 30
raspberries, aroma 30
refrigerated transport 127–28
refrigeration 115–17
relative humidity (RH) 82, 250
resistance thermometers 248

respiration 33, 36
  biochemistry of 47–53
  effect of modified atmosphere on 100–102
  physiology of 37–39
respiration equation 97
respiration rate 231, 233
respiratory quotient *see* RQ
*Rhizopus* 144, 147, 152, 155
*R. stolonifer* 156
ripening 34, 35–46, 54–56, 169–70
  controlled 205–209
  delayed by low temperature 61–63
  role of ethylene in 106–07
  *see also* maturation
room cooling 69
root-rot 156
roses
  controlled atmosphere 99
  harvesting 170
  packaging and transport 93, 95
rotting
  at high humidity 90
  reduction of 102
RQ (respiratory quotient) 50–51
Safe Quality Food *see* SQF 2000
sanitation 126–27
saturated air 82
saturation vapour pressure 250
scale insects 194
*Sclerotinia* 144, 146, 156
scurvy 1, 163
sec-butylamine (2-AB) 154
seeds and pods 57
senescence 34, 45
senescent blotch 138
senescent breakdown 138
seven-eighths cooling time 68, 69
shape 161, 172–73
shrinkage and weight loss 126
silver ions 110
silver thiosulphate *see* STS
size
  fruit 161, 172–73
  packaging 220–21
skin blemishes 137–40, 162
sodium *o*-phenylphenate *see* SOPP
soft rot 152, 153
soft scald 138
soluble solids concentration (SSC) 212
SOPP (sodium *o*-phenylphenate) 149, 152, 155
sorting of produce 124

sour rot 153, 154, 156
specific humidity 250
spinach, vitamins in 27, 28
sprouted seeds 56–57
sprout inhibitors 192–93, 201
SQF 2000 (Safe Quality Food) 182
stacking 125
standards
   irradiated food 201
   quality 180–82
starch and sugar 23, 55, 65,
   174–75
   degradation 48
stem-end rots 139, 145, 146,
   149, 152, 153
   in mangoes *colour plate*
stomata 87–88
storage 167
   atmosphere 97–112
   controlled temperature and
      atmosphere 52–53
   environment control 11
   low temperature 66
   methods 113–17
   recommendations 230–40
   technology of 113–29
storage spot *colour plate*
stowage 224
strawberries
   disorders 142, 153
   effect of modified atmosphere
      102, 103
   grey mould in 147, 148
   injury 219
   radiation 202–203
   respiration 233
   storage life 232
   transport 98
   vitamins in 27
STS (silver thiosulphate) 110
sucrose pulsing 211
sugar and starch 22–23, 55, 65,
   174–75
   degradation 48
sulphur dioxide 149, 150, 155
summer squash, injury 219
sunburn scald 138
superficial scald *colour plate*, 137,
   138–39, 139
   control 202–204
surface area to volume ratio in
   water loss 86
surface nature of produce 87–88
surfactants 95
sweet corn 164
   vacuum cooling 74
   vitamins in 28
sweet potatoes 191
   disorders 153

storage 65, 115, 232
   vitamins in 27, 28
tartaric acid 25
TBZ *see* thiabendazole
TCA (tricarboxylic acid) cycle
   47, 48, 49
temperature
   effect on produce 60–76,
      130–35, 136–37
   measurement 246–49
   placement of thermometer 249
   storage 231–33
   storage of ornamentals 239
   *see also* cold treatment; cooling;
      heat treatment
temperature control in stores
   11, 123–24
tenderometer 170
tetrazine 226
tetrazines 110
thermal conductivity gas
   chromatography 254
thermocouples 248
thermographs 247
thermometers 246–49
   calibration 249
thiabendazole (TBZ) *colour plate*,
   150, 152
thin-film polymer hygrometers
   252
thiophanate methyl 152
tight-fill packing 223
tomatoes
   chilling injury 131
   composition 31
   disorders 142
   ethylene in 40
   injury 219
   maturity 44
   mutants 45–46, 53
   radiation 202–203
   ripening *colour plate*, 44, 46,
      142, 207, 209
   ripening physiochemical
      changes 35
   storage life 232
transpiration coefficients 86
   vitamins in 27, 28
TQM (Total Quality
   Management) 181–82
transit rot 156
transmission spectrophotometers
   174, 175
transpiration 33, 81, 165–66
transpiration coefficients for
   horticultural produce 86
transport and handling 166–67
treatments, postharvest 190–93,
   196–97, 200–12

triazoles 150, 154–55
tricarboxylic acid cycle *see* TCA
   cycle
trickle degreening 210
trickle system 206, 208
tropical and sub-tropical fruit,
   storage 237
turgor potential 79
underground vegetables, storage
   conditions 234, 236
uptake preservation 212
vacuum cooling 73–74
vapour heat 197, 200
vapour pressure deficit (VPD)
   82–85
vegetable fruits 15
   storage conditions 234
vegetables
   chemical composition 21–32
   defined 15–16
   maturation 56–57
   physiology and biochemistry
      33–59
   structure 15–17, 18
world production 3–5
vibration injury 216, 217–18
vine fruit storage 236
vitamin A 26, 27
vitamin C 1, 25–26, 27
   loss 78
vitamins and minerals 25–28
volatiles 29–30, 56
volume packing 223
VPD *see* vapour pressure deficit
washing 190
wastage control 146–57
water 21–22
water core 138
water loss 207
   factors affecting 86–89
   and humidity 77–96
   in packages 226
   reducing 89–93
water potential 79–80, 82–83
waxing *colour plate*, 190–91
weight loss
   effect of packaging on 220–21
   and shrinkage 126
wet and dry bulb hygrometers 251
whisker mould 153
wilting 81, 165–66
   of cut flowers and foliage 94
woolliness in peach 133, 139
x-ray tomography 178
xylem water potential 80
zucchini, vacuum cooling 74